British Railways Pocket Book No. 4

ELECTRIC MULTIPLE UNITS

TWENTY SIXTH EDITION

The C(
Electric Multip
the national rail
the major UK

Robert Pritchard

ISBN 978 1909 431 01 0

CONTENTS

PROVISION OF INFORMATION

This book has been compiled with care to be as accurate as possible, but in some cases information is not officially available and the publisher cannot be held responsible for any errors or omissions. We would like to thank the companies and individuals which have been co-operative in supplying information to us. The authors of this series of books are always pleased to receive notification from readers of any inaccuracies readers may find in the series, to enhance future editions. Please send comments to:

Robert Pritchard, Platform 5 Publishing Ltd, 3 Wyvern House, Sark Road, Sheffield, S2 4HG, England.

e-mail: robert@platform5.com   **Tel:** 0114 255 2625   **Fax:** 0114 255 2471

This book is updated to information received by 8 October 2012.

UPDATES

This book is updated to the Stock Changes given in **Today's Railways UK 131** (November 2012). Readers are therefore advised to update this book from the official Platform 5 Stock Changes published every month in **Today's Railways UK** magazine, starting with issue 132.

The Platform 5 magazine **Today's Railways UK** contains news and rolling stock information on the railways of Britain and Ireland and is published on the second Monday of every month. For further details of **Today's Railways UK**, please see the advertisement on the back cover of this book.

Front cover photograph: South West Trains 444 022 passes Holes Bay, Poole, on 23/07/12 working the 12.03 Weymouth–London Waterloo.   **Andrew Mist**

BRITAIN'S RAILWAY SYSTEM

INFRASTRUCTURE & OPERATION

Britain's national railway infrastructure is owned by a "not for dividend" company, Network Rail (NR). Many stations and maintenance depots are leased to and operated by Train Operating Companies (TOCs), but some larger stations remain under Network Rail control. The only exception is the infrastructure on the Isle of Wight, which is nationally owned and is leased to South West Trains.

Trains are operated by TOCs over Network Rail, regulated by access agreements between the parties involved. In general, TOCs are responsible for the provision and maintenance of the locomotives, rolling stock and staff necessary for the direct operation of services, whilst NR is responsible for the provision and maintenance of the infrastructure and also for staff to regulate the operation of services.

At the time of going to press the lettering of new franchises was on hold pending a review on the franchise system ordered by the Government following the cancelling of the award of the West Coast franchise to First Group in October 2012.

DOMESTIC PASSENGER TRAIN OPERATORS

The large majority of passenger trains are operated by the TOCs on fixed term franchises. Franchise expiry dates are shown in the list of franchisees below:

Franchise	Franchisee	Trading Name
Chiltern Railways	Deutsche Bahn (Arriva) (until 31 December 2021)	Chiltern Railways
Cross-Country[1]	Deutsche Bahn (Arriva) (until 11 November 2013)	CrossCountry
East Midlands[2]	Stagecoach Group plc (until 31 March 2015)	East Midlands Trains
Greater Western	First Group plc (until 31 March 2013)	First Great Western
Greater Anglia[3]	Abellio (until 19 July 2014)	Greater Anglia
Integrated Kent	GoVia Ltd (Go-Ahead/Keolis) (until 31 March 2014)	Southeastern
InterCity East Coast[4]		East Coast
InterCity West Coast[5]	Virgin Rail Group Ltd. (until 8 December 2012)	Virgin Trains
London Rail[6]	MTR/Deutsche Bahn (until 31 March 2014)	London Overground
LTS Rail	National Express Group plc (until 26 May 2013)	c2c
Merseyrail Electrics[7]	Serco/Abellio (until 19 July 2028)	Merseyrail
Northern Rail[8]	Serco/Abellio (until 1 April 2014)	Northern

ScotRail	First Group plc (until 8 November 2014)	ScotRail
South Central[9]	Govia Ltd (Go-Ahead/Keolis) (until 25 July 2015)	Southern
South Western[10]	Stagecoach Group plc (until 3 February 2014)	South West Trains
Thameslink/Great Northern[11]	First Group plc (until September 2013)	First Capital Connect
Trans-Pennine Express[12]	First Group/Keolis (until 1 April 2014)	TransPennine Express
Wales & Borders	Deutsche Bahn (Arriva) (until 6 December 2018)	Arriva Trains Wales
West Midlands[13]	Govia Ltd (Go-Ahead/Keolis) (until 11 November 2013)	London Midland

Notes on franchise end dates:

[1] Awarded for six years to 2013 with an extension for a further two years and five months to 31 March 2016 if performance targets are met.

[2] Likely to be extended for four weeks to 28 April 2013 owing to engineering works at Reading over the Easter 2013 period.

[3] A new short-term Greater Anglia franchise commenced on 5 February 2012, operated by Abellio. This is due to run until July 2014 when a new 15-year franchise is due to start.

[4] Currently run on an interim basis by DfT management company Directly Operated Railways (trading as East Coast). This arrangement is due to continue until at least December 2013.

[5] The Virgin Trains franchise was due to end in December 2012 but the award of the new franchise to First Group was cancelled in October 2012 and at the time of writing it had not been decided whether Virgin Trains would should continue running the franchise or whether it should be taken over by the Government (Directly Operated Railways) on an interim basis.

[6] The London Rail Concession is different from other rail franchises, as fares and service levels are set by Transport for London instead of the DfT.

[7] Under the control of Merseytravel PTE instead of the DfT. Franchise reviewed every five years to fit in with the Merseyside Local Transport Plan.

[8] Franchise extended in 2011 to end between April 2014 and March 2015 to align with the end of the Trans-Pennine Express franchise.

[9] Upon termination of the Southern franchise in July 2015 it is to be combined with the Thameslink/Great Northern franchise.

[10] Awarded for seven years to 2014 with an extension for a further three years to 3 February 2017 if performance targets are met.

[11] Awarded for six years to 2012 with an extension for up to a further three years to 31 March 2015 if performance targets were met. This was cut by 18 months in 2011 so the franchise would finish in September 2013. The new franchise will be combined with Southern from July 2015.

[12] Originally awarded for eight years to 2012 with a possible extension for five years to 31 January 2017. The end date was revised in 2011 to be aligned with the end of the Northern franchise – the present contract will finish between April 2014 and March 2015.

[13] Awarded for six years to 2013 with an extension for a further two years to 19 September 2015 if performance targets are met.

All new franchises officially start at 02.00 on the first day, although the last dates shown above are the last full day of operation.

Where termination dates are dependent on performance targets being met, the earliest possible termination date is given. However, with Merseyrail the end date is based on the maximum franchise length.

The following operators run non-franchised services only:

Operator	Trading Name	Route
BAA	Heathrow Express	London Paddington–Heathrow Airport
First Hull Trains	First Hull Trains	London King's Cross–Hull
Grand Central	Grand Central	London King's Cross–Sunderland/ Bradford Interchange
North Yorkshire Moors Railway Enterprises	North Yorkshire Moors Railway	Pickering–Grosmont–Whitby/ Battersby
West Coast Railway Company	West Coast Railway Company	Birmingham–Stratford-upon-Avon* Fort William–Mallaig* York–Wakefield–York–Scarborough*

* Special summer services only.

INTERNATIONAL PASSENGER OPERATIONS

Eurostar (UK) operates passenger services between the UK and mainland Europe, jointly with the national operators of France (SNCF) and Belgium (SNCB/NMBS). Eurostar International is owned by three shareholders: SNCF (55%), London & Continental Railways (40%) and SNCB (5%).

In addition, a service for the conveyance of accompanied road vehicles through the Channel Tunnel is provided by the tunnel operating company, Eurotunnel.

FREIGHT TRAIN OPERATIONS

The following operators operate freight services or empty passenger stock workings under "Open Access" arrangements:

Colas Rail
DB Schenker Rail (UK)
Devon & Cornwall Railways
Direct Rail Services (DRS)
Freightliner
GB Railfreight (owned by Eurotunnel)
West Coast Railway Company

INTRODUCTION

EMU CLASSES

Principal details and dimensions are quoted for each class in metric and/or imperial units as considered appropriate bearing in mind common UK usage.

All dimensions and weights are quoted for vehicles in an "as new" condition with all necessary supplies on board. Dimensions are quoted in the order length x overall width. All lengths quoted are over buffers or couplers as appropriate. Where two lengths are quoted, the first refers to outer vehicles in a set and the second to inner vehicles. All weights are shown as metric tonnes (t = tonnes).

Bogie Types are quoted in the format motored/non-motored (eg BP20/BT13 denotes BP20 motored bogies and BT non-motored bogies).

Unless noted to the contrary, all vehicles listed have bar couplers at non-driving ends.

Vehicles ordered under the auspices of BR were allocated a Lot (batch) number when ordered and these are quoted in class headings and sub-headings. Vehicles ordered since 1995 have no Lot Numbers, but the manufacturer and location that they were built is given.

NUMERICAL LISTINGS

25 kV AC 50 Hz overhead Electric Multiple Units (EMUs) and dual voltage EMUs are listed in numerical order of set numbers. Individual "loose" vehicles are listed in numerical order after vehicles formed into fixed formations.

750 V DC third rail EMUs are listed in numerical order of class number, then in numerical order of set number. Some of these use the former Southern Region four-digit set numbers. These are derived from theoretical six digit set numbers which are the four-digit set number prefixed by the first two numbers of the class.

Where sets or vehicles have been renumbered in recent years, former numbering detail is shown alongside current detail. Each entry is laid out as in the following example:

Set No.	Detail	Livery	Owner	Operator	Allocation	Formation
320314	*	**SR**	E	*SR*	GW	77912 63034 77934

Detail Differences. Only detail differences which currently affect the areas and types of train which vehicles may work are shown. All other detail differences are excluded. Where such differences occur within a class or part class, these are shown alongside the individual set or vehicle number. Meaning of abbreviations are detailed in individual class headings.

Set Formations. Set formations shown are those normally maintained. Readers should note some set formations might be temporarily varied from time to time to suit maintenance and/or operational requirements. Vehicles shown as "Spare" are not formed in any regular set formation.

Codes. Codes are used to denote the livery, owner, operator and depot of each unit. Details of these will be found in Section 7 of this book. Where a unit or spare car is off-lease, the operator column will be left blank.

Names. Only names carried with official sanction are listed. Names are shown in UPPER/lower case characters as actually shown on the name carried on the vehicle(s). Unless otherwise shown, complete units are regarded as named rather than just the individual car(s) which carry the name.

GENERAL INFORMATION

CLASSIFICATION AND NUMBERING

25 kV AC 50 Hz overhead and "Versatile" EMUs are classified in the series 300–399.

750 V DC third rail EMUs are classified in the series 400–599.

Service units are classified in the series 900–949.

EMU individual cars are numbered in the series 61000–78999, except for vehicles used on the Isle of Wight – which are numbered in a separate series, and the Class 378s, 380s and 395s, which take up 38xxx and 39xxx series'.

Any vehicle constructed or converted to replace another vehicle following accident damage and carrying the same number as the original vehicle is denoted by the suffix[II] in this publication.

OPERATING CODES

These codes are used by train operating company staff to describe the various different types of vehicles and normally appear on data panels on the inner (ie non driving) ends of vehicles.

A "B" prefix indicates a battery vehicle.
A "P" prefix indicates a trailer vehicle on which is mounted the pantograph, instead of the default case where the pantograph is mounted on a motor vehicle.

The first part of the code describes whether or not the car has a motor or a driving cab as follows:

DM	Driving motor	M	Motor
DT	Driving trailer	T	Trailer

The next letter is a "B" for cars with a brake compartment.
This is followed by the saloon details:

F	First	S	Standard
C	Composite		

The next letter denotes the style of accommodation as follows:

O	Open	K	Side compartment with lavatory
so	Semi-open (part compartments, part open). All other vehicles are		
	assumed to consist solely of open saloons.		

Finally vehicles with a buffet are suffixed RB or RMB for a miniature buffet.

Where two vehicles of the same type are formed within the same unit, the above codes may be suffixed by (A) and (B) to differentiate between the vehicles.

A composite is a vehicle containing both First and Standard Class accommodation, whilst a brake vehicle is a vehicle containing separate specific accommodation for the conductor.

Special Note: Where vehicles have been declassified, the correct operating code which describes the actual vehicle layout is quoted in this publication.

The following codes are used to denote special types of vehicle:

DMLF	Driving Motor Lounge First
DMLV	Driving Motor Luggage Van
MBRBS	Motor buffet standard with luggage space and guard's compartment.
TFH	Trailer First with Handbrake

BUILD DETAILS

Lot Numbers
Vehicles ordered under the auspices of BR were allocated a Lot (batch) number when ordered and these are quoted in class headings and sub-headings.

ACCOMMODATION

The information given in class headings and sub-headings is in the form F/S nT (or TD) nW. For example 12/54 1T 1W denotes 12 First Class and 54 Standard Class seats, one toilet and one space for a wheelchair. A number in brackets (i.e. (2)) denotes tip-up seats (in addition to the fixed seats). Tip-up seats in vestibules do not count. The seating layout of open saloons is shown as 2+1, 2+2 or 3+2 as the case may be. Where units have first class accommodation as well as standard and the layout is different for each class then these are shown separately prefixed by "1:" and "2:". Compartments are three seats a side in First Class and mostly four a side in Standard Class in EMUs. TD denotes a toilet suitable for use by a disabled person.

ABBREVIATIONS

The following standard abbreviations are used in class headings and also throughout this publication:

AC	Alternating Current.	kW	kilowatts.
BR	British Railways.	LT	London Transport.
BSI	Bergische Stahl Industrie.	LUL	London Underground Limited.
DC	Direct Current.	m	metres.
EMU	Electric Multiple Unit.	mph	miles per hour.
Hz	Hertz.	SR	BR Southern Region.
kN	kilonewtons.	t	tonnes.
km/h	kilometres per hour.	V	volts.

1. 25 kV AC 50 Hz OVERHEAD & DUAL VOLTAGE UNITS

Except where otherwise stated, all units in this section operate on 25 kV AC 50 Hz overhead only.

CLASS 313   BREL YORK

Inner suburban units.

Formation: DMSO–PTSO–BDMSO or DMSO–TSO–BDMSO.
Systems: 25 kV AC overhead/750 V DC third rail.
Construction: Steel underframe, aluminium alloy body and roof.
Traction Motors: Four GEC G310AZ of 82.125 kW.
Wheel Arrangement: Bo-Bo + 2-2 + Bo-Bo.
Braking: Disc & rheostatic.   **Dimensions:** 20.33/20.18 x 2.82 m.
Bogies: BX1.   **Couplers:** Tightlock.
Gangways: Within unit + end doors. **Control System:** Camshaft.
Doors: Sliding.   **Maximum Speed:** 75 mph.
Seating Layouts: 313/0: Refurbished with high back seats (3+2 facing).
313/1: low back seats (3+2 facing).
313/2: refurbished with 2+2 mainly facing high-back seating.
Multiple Working: Within class.

DMSO. Lot No. 30879 1976–77. –/74. 36.0 t.
PTSO. Lot No. 30880 1976–77. –/83. 31.0 t.
BDMSO. Lot No. 30885 1976–77. –/74. 37.5 t.

Class 313/0. Standard Design.

313018	**FU**	E	*FC*	HE	62546	71230	62610
313024	**FU**	E	*FC*	HE	62552	71236	62616
313025	**FU**	E	*FC*	HE	62553	71237	62617
313026	**FU**	E	*FC*	HE	62554	71238	62618
313027	**FU**	E	*FC*	HE	62555	71239	62619
313028	**FU**	E	*FC*	HE	62556	71240	62620
313029	**FU**	E	*FC*	HE	62557	71241	62621
313030	**FU**	E	*FC*	HE	62558	71242	62622
313031	**FU**	E	*FC*	HE	62559	71243	62623
313032	**FU**	E	*FC*	HE	62560	71244	62643
313033	**FU**	E	*FC*	HE	62561	71245	62625
313035	**FU**	E	*FC*	HE	62563	71247	62627
313036	**FU**	E	*FC*	HE	62564	71248	62628
313037	**FU**	E	*FC*	HE	62565	71249	62629
313038	**FU**	E	*FC*	HE	62566	71250	62630
313039	**FU**	E	*FC*	HE	62567	71251	62631
313040	**FU**	E	*FC*	HE	62568	71252	62632
313041	**FU**	E	*FC*	HE	62569	71253	62633
313042	**FU**	E	*FC*	HE	62570	71254	62634
313043	**FU**	E	*FC*	HE	62571	71255	62635
313044	**FU**	E	*FC*	HE	62572	71256	62636

313045	**FU**	E	*FC*	HE	62573	71257	62637
313046	**FU**	E	*FC*	HE	62574	71258	62638
313047	**FU**	E	*FC*	HE	62575	71259	62639
313048	**FU**	E	*FC*	HE	62576	71260	62640
313049	**FU**	E	*FC*	HE	62577	71261	62641
313050	**FU**	E	*FC*	HE	62578	71262	62649
313051	**FU**	E	*FC*	HE	62579	71263	62624
313052	**FU**	E	*FC*	HE	62580	71264	62644
313053	**FU**	E	*FC*	HE	62581	71265	62645
313054	**FU**	E	*FC*	HE	62582	71266	62646
313055	**FU**	E	*FC*	HE	62583	71267	62647
313056	**FU**	E	*FC*	HE	62584	71268	62648
313057	**FU**	E	*FC*	HE	62585	71269	62642
313058	**FU**	E	*FC*	HE	62586	71270	62650
313059	**FU**	E	*FC*	HE	62587	71271	62651
313060	**FU**	E	*FC*	HE	62588	71272	62652
313061	**FU**	E	*FC*	HE	62589	71273	62653
313062	**FU**	E	*FC*	HE	62590	71274	62654
313063	**FU**	E	*FC*	HE	62591	71275	62655
313064	**FU**	E	*FC*	HE	62592	71276	62656

Name (carried on PTSO): 313054   Captain William Leefe Robinson V.C.

Class 313/1. Former London Overground units. Details as Class 313/0.

313121	**SL**	BN		WB	62549	71233	62613
313122	**FU**	E	*FC*	HE	62550	71234	62614
313123	**FU**	E	*FC*	HE	62551	71235	62615
313134	**FU**	E	*FC*	HE	62562	71246	62626

Name (carried on PTSO): 313134   City of London

Class 313/2. Southern units. Units refurbished for Southern for Brighton Coastway services. 750V DC only (pantographs removed).

DMSO. Lot No. 30879 1976–77. –/64. 36.0 t.
TSO. Lot No. 30880 1976–77. –/68.   t.
BDMSO. Lot No. 30885 1976–77. –/64. 37.5 t.

313201	(313101)	**SN**	BN	*SN*	BI	62529	71213	62593
313202	(313102)	**SN**	BN	*SN*	BI	62530	71214	62594
313203	(313103)	**SN**	BN	*SN*	BI	62531	71215	62595
313204	(313104)	**SN**	BN	*SN*	BI	62532	71216	62596
313205	(313105)	**SN**	BN	*SN*	BI	62533	71217	62597
313206	(313106)	**SN**	BN	*SN*	BI	62534	71218	62598
313207	(313107)	**SN**	BN	*SN*	BI	62535	71219	62599
313208	(313108)	**SN**	BN	*SN*	BI	62536	71220	62600
313209	(313109)	**SN**	BN	*SN*	BI	62537	71221	62601
313210	(313110)	**SN**	BN	*SN*	BI	62538	71222	62602
313211	(313111)	**SN**	BN	*SN*	BI	62539	71223	62603
313212	(313112)	**SN**	BN	*SN*	BI	62540	71224	62604
313213	(313113)	**SN**	BN	*SN*	BI	62541	71225	62605
313214	(313114)	**SN**	BN	*SN*	BI	62542	71226	62606
313215	(313115)	**SN**	BN	*SN*	BI	62543	71227	62607

313216	(313116)	**SN**	BN	*SN*	BI	62544	71228	62608
313217	(313117)	**SN**	BN	*SN*	BI	62545	71229	62609
313219	(313119)	**SN**	BN	*SN*	BI	62547	71231	62611
313220	(313120)	**SN**	BN	*SN*	BI	62548	71232	62612

CLASS 314   BREL YORK

Inner suburban units.

Formation: DMSO–PTSO–DMSO.
Construction: Steel underframe, aluminium alloy body and roof.
Traction Motors: Four GEC G310AZ (* Brush TM61-53) of 82.125 kW.
Wheel Arrangement: Bo-Bo + 2-2 + Bo-Bo.
Braking: Disc & rheostatic.   **Dimensions:** 20.33/20.18 x 2.82 m.
Bogies: BX1.   **Couplers:** Tightlock.
Gangways: Within unit + end doors. **Control System:** Thyristor.
Doors: Sliding.   **Maximum Speed:** 70 mph.
Seating Layout: 3+2 low-back facing.
Multiple Working: Within class and with Class 315.

DMSO. Lot No. 30912 1979. –/68. 34.5 t.
64588. **DMSO.** Lot No. 30908 1978–80. Rebuilt Railcare Glasgow 1996 from Class 507 No. 64426. The original 64588 was scrapped. –/74. 34.5 t.
PTSO. Lot No. 30913 1979. –/76. 33.0 t.

314201	*	**SC**	A	*SR*	GW	64583	71450	64584
314202	*	**SC**	A	*SR*	GW	64585	71451	64586
314203	*	**SR**	A	*SR*	GW	64587	71452	64588
314204	*	**SR**	A	*SR*	GW	64589	71453	64590
314205	*	**SC**	A	*SR*	GW	64591	71454	64592
314206	*	**SC**	A	*SR*	GW	64593	71455	64594
314207		**SC**	A	*SR*	GW	64595	71456	64596
314208		**SC**	A	*SR*	GW	64597	71457	64598
314209		**SC**	A	*SR*	GW	64599	71458	64600
314210		**SC**	A	*SR*	GW	64601	71459	64602
314211		**SR**	A	*SR*	GW	64603	71460	64604
314212		**SR**	A	*SR*	GW	64605	71461	64606
314213		**SC**	A	*SR*	GW	64607	71462	64608
314214		**SC**	A	*SR*	GW	64609	71463	64610
314215		**SC**	A	*SR*	GW	64611	71464	64612
314216		**SC**	A	*SR*	GW	64613	71465	64614

CLASS 315   BREL YORK

Inner suburban units.

Formation: DMSO–TSO–PTSO–DMSO.
Construction: Steel underframe, aluminium alloy body and roof.
Traction Motors: Four Brush TM61-53 (* GEC G310AZ) of 82.125 kW.
Wheel Arrangement: Bo-Bo + 2-2 + 2-2 + Bo-Bo.
Braking: Disc & rheostatic.   **Dimensions:** 20.33/20.18 x 2.82 m.
Bogies: BX1.   **Couplers:** Tightlock.

Gangways: Within unit + end doors. **Control System:** Thyristor.
Doors: Sliding.   **Maximum Speed:** 75 mph.
Seating Layout: 3+2 low-back facing.
Multiple Working: Within class and with Class 314.

DMSO. Lot No. 30902 1980–81. –/74. 35.0 t.
TSO. Lot No. 30904 1980–81. –/86. 25.5 t.
PTSO. Lot No. 30903 1980–81. –/84. 32.0 t.

315801		1	E	EA	IL	64461	71281	71389	64462
315802		1	E	EA	IL	64463	71282	71390	64464
315803		1	E	EA	IL	64465	71283	71391	64466
315804	GA	E	EA	IL	64467	71284	71392	64468
315805	1	E	EA	IL	64469	71285	71393	64470
315806	GA	E	EA	IL	64471	71286	71394	64472
315807	1	E	EA	IL	64473	71287	71395	64474
315808	1	E	EA	IL	64475	71288	71396	64476
315809	GA	E	EA	IL	64477	71289	71397	64478
315810	1	E	EA	IL	64479	71290	71398	64480
315811	1	E	EA	IL	64481	71291	71399	64482
315812	1	E	EA	IL	64483	71292	71400	64484
315813	1	E	EA	IL	64485	71293	71401	64486
315814	1	E	EA	IL	64487	71294	71402	64488
315815	1	E	EA	IL	64489	71295	71403	64490
315816	1	E	EA	IL	64491	71296	71404	64492
315817	1	E	EA	IL	64493	71297	71405	64494
315818	1	E	EA	IL	64495	71298	71406	64496
315819	1	E	EA	IL	64497	71299	71407	64498
315820	1	E	EA	IL	64499	71300	71408	64500
315821	1	E	EA	IL	64501	71301	71409	64502
315822	1	E	EA	IL	64503	71302	71410	64504
315823	1	E	EA	IL	64505	71303	71411	64506
315824	1	E	EA	IL	64507	71304	71412	64508
315825	1	E	EA	IL	64509	71305	71413	64510
315826	1	E	EA	IL	64511	71306	71414	64512
315827	1	E	EA	IL	64513	71307	71415	64514
315828	1	E	EA	IL	64515	71308	71416	64516
315829	1	E	EA	IL	64517	71309	71417	64518
315830	1	E	EA	IL	64519	71310	71418	64520
315831	1	E	EA	IL	64521	71311	71419	64522
315832	1	E	EA	IL	64523	71312	71420	64524
315833	1	E	EA	IL	64525	71313	71421	64526
315834	1	E	EA	IL	64527	71314	71422	64528
315835	1	E	EA	IL	64529	71315	71423	64530
315836	1	E	EA	IL	64531	71316	71424	64532
315837	1	E	EA	IL	64533	71317	71425	64534
315838	1	E	EA	IL	64535	71318	71426	64536
315839	1	E	EA	IL	64537	71319	71427	64538
315840	1	E	EA	IL	64539	71320	71428	64540
315841	1	E	EA	IL	64541	71321	71429	64542
315842	*	1	E	EA	IL	64543	71322	71430	64544
315843	*	1	E	EA	IL	64545	71323	71431	64546

315844	*	1	E	*EA*	IL	64547	71324	71432	64548
315845	*	1	E	*EA*	IL	64549	71325	71433	64550
315846	*	1	E	*EA*	IL	64551	71326	71434	64552
315847	*	1	E	*EA*	IL	64553	71327	71435	64554
315848	*	1	E	*EA*	IL	64555	71328	71436	64556
315849	*	1	E	*EA*	IL	64557	71329	71437	64558
315850	*	1	E	*EA*	IL	64559	71330	71438	64560
315851	*	1	E	*EA*	IL	64561	71331	71439	64562
315852	*	1	E	*EA*	IL	64563	71332	71440	64564
315853	*	1	E	*EA*	IL	64565	71333	71441	64566
315854	*	1	E	*EA*	IL	64567	71334	71442	64568
315855	*	1	E	*EA*	IL	64569	71335	71443	64570
315856	*	1	E	*EA*	IL	64571	71336	71444	64572
315857	*	1	E	*EA*	IL	64573	71337	71445	64574
315858	*	1	E	*EA*	IL	64575	71338	71446	64576
315859	*	1	E	*EA*	IL	64577	71339	71447	64578
315860	*	1	E	*EA*	IL	64579	71340	71448	64580
315861	*	1	E	*EA*	IL	64581	71341	71449	64582

Names (carried on DMSO):

315817	Transport for London
315829	London Borough of Havering Celebrating 40 years
315845	Herbie Woodward
315857	Stratford Connections

CLASS 317   BREL YORK/DERBY

Outer suburban units.

Formation: Various, see sub-class headings.
Construction: Steel.
Traction Motors: Four GEC G315BZ of 247.5 kW.
Wheel Arrangement: 2-2 + Bo-Bo + 2-2 + 2-2.
Braking: Disc.   **Dimensions:** 20.13/20.18 x 2.82 m.
Bogies: BP20 (MSO), BT13 (others).   **Couplers:** Tightlock.
Gangways: Throughout   **Control System:** Thyristor.
Doors: Sliding.   **Maximum Speed:** 100 mph.
Seating Layout: Various, see sub-class headings.
Multiple Working: Within class & with Classes 318, 319, 320, 321, 322 and 323.

Class 317/1. Pressure ventilated.

Formation: DTSO–MSO–TCO–DTSO.
Seating Layout: 1: 2+2 facing, 2: 3+2 facing.

DTSO(A) Lot No. 30955 York 1981–82. –/74. 29.5 t.
MSO. Lot No. 30958 York 1981–82. –/79. 49.0 t.
TCO. Lot No. 30957 Derby 1981–82. 22/46 2T. 29.0 t.
DTSO(B) Lot No. 30956 York 1981–82. –/71. 29.5 t.

317337	**FU**	A	*FC*	HE	77036	62671	71613	77084
317338	**FU**	A	*FC*	HE	77037	62698	71614	77085
317339	**FU**	A	*FC*	HE	77038	62699	71615	77086

317340	**FU**	A	*FC*	HE	77039	62700	71616	77087
317341	**FU**	A	*FC*	HE	77040	62701	71617	77088
317342	**FU**	A	*FC*	HE	77041	62702	71618	77089
317343	**FU**	A	*FC*	HE	77042	62703	71619	77090
317344	**FU**	A	*FC*	HE	77029	62690	71620	77091
317345	**FU**	A	*FC*	HE	77044	62705	71621	77092
317346	**FU**	A	*FC*	HE	77045	62706	71622	77093
317347	**FU**	A	*FC*	HE	77046	62707	71623	77094
317348	**FU**	A	*FC*	HE	77047	62708	71624	77095

Names (carried on TCO):

317345 Driver John Webb | 317348 Richard A Jenner

Class 317/5. Pressure ventilated. Units renumbered from Class 317/1 in 2005 for West Anglia Metro services. Refurbished with new upholstery and Passenger Information Systems. Details as Class 317/1.

Note: The original DTSO 77048 was written off after the Cricklewood accident of 1983. A replacement vehicle was built (at Wolverton) in 1987 and given the same number.

317501	**NX**	A	*EA*	IL	77024	62661	71577	77048
317502	**NX**	A	*EA*	IL	77001	62662	71578	77049
317503	**NX**	A	*EA*	IL	77002	62663	71579	77050
317504	**NX**	A	*EA*	IL	77003	62664	71580	77051
317505	**NX**	A	*EA*	IL	77004	62665	71581	77052
317506	**NX**	A	*EA*	IL	77005	62666	71582	77053
317507	**NX**	A	*EA*	IL	77006	62667	71583	77054
317508	**NX**	A	*EA*	IL	77010	62697	71587	77058
317509	**NX**	A	*EA*	IL	77011	62672	71588	77059
317510	**NX**	A	*EA*	IL	77012	62673	71589	77060
317511	**NC**	A	*EA*	IL	77014	62675	71591	77062
317512	**NC**	A	*EA*	IL	77015	62676	71592	77063
317513	**NX**	A	*EA*	IL	77016	62677	71593	77064
317514	**NX**	A	*EA*	IL	77017	62678	71594	77065
317515	**NX**	A	*EA*	IL	77019	62680	71596	77067

Name (carried on TCO):

317507 University of Cambridge 800 Years 1209–2009

Class 317/6. Convection heating. Units converted from Class 317/2 by Railcare, Wolverton 1998–99 with new seating layouts.

Formation: DTSO–MSO–TSO–DTCO.
Seating Layout: 2+2 facing.

77200–219. DTSO. Lot No. 30994 York 1985–86. –/64. 29.5 t.
77280–283. DTSO. Lot No. 31007 York 1987. –/64. 29.5 t.
62846–865. MSO. Lot No. 30996 York 1985–86. –/70. 49.0 t.
62886–889. MSO. Lot No. 31009 York 1987. –/70. 49.0 t.
71734–753. TSO. Lot No. 30997 York 1985–86. –/62 2T. 29.0 t.
71762–765. TSO. Lot No. 31010 York 1987. –/62 2T. 29.0 t.
77220–239. DTCO. Lot No. 30995 York 1985–86. 24/48. 29.5 t.
77284–287. DTCO. Lot No. 31008 York 1987. 24/48. 29.5 t.

317649	**NC**	A	*EA*	IL	77200	62846	71734	77220
317650	**NC**	A	*EA*	IL	77201	62847	71735	77221
317651	**NC**	A	*EA*	IL	77202	62848	71736	77222
317652	**NC**	A	*EA*	IL	77203	62849	71739	77223
317653	**NC**	A	*EA*	IL	77204	62850	71738	77224
317654	**NC**	A	*EA*	IL	77205	62851	71737	77225
317655	**NC**	A	*EA*	IL	77206	62852	71740	77226
317656	**NC**	A	*EA*	IL	77207	62853	71742	77227
317657	**NC**	A	*EA*	IL	77208	62854	71741	77228
317658	**NC**	A	*EA*	IL	77209	62855	71743	77229
317659	**GA**	A	*EA*	IL	77210	62856	71744	77230
317660	**GA**	A	*EA*	IL	77211	62857	71745	77231
317661	**GA**	A	*EA*	IL	77212	62858	71746	77232
317662	**GA**	A	*EA*	IL	77213	62859	71747	77233
317663	**1**	A	*EA*	IL	77214	62860	71748	77234
317664	**1**	A	*EA*	IL	77215	62861	71749	77235
317665	**1**	A	*EA*	IL	77216	62862	71750	77236
317666	**NC**	A	*EA*	IL	77217	62863	71752	77237
317667	**1**	A	*EA*	IL	77218	62864	71751	77238
317668	**1**	A	*EA*	IL	77219	62865	71753	77239
317669	**NC**	A	*EA*	IL	77280	62886	71762	77284
317670	**1**	A	*EA*	IL	77281	62887	71763	77285
317671	**NC**	A	*EA*	IL	77282	62888	71764	77286
317672	**GA**	A	*EA*	IL	77283	62889	71765	77287

Name (carried on DTCO): 317654 Richard Wells

Class 317/7. Units converted from Class 317/1 by Railcare, Wolverton 2000 for Stansted Express services between London Liverpool Street and Stansted. Air conditioning. Fitted with luggage stacks. Displaced from Stansted services in 2011 by Class 379s and all now in store. 317 722 is to become a pilot unit for a re-engineering project led by Angel Trains.

Formation: DTSO–MSO–TSO–DTCO.
Seating Layout: 1: 2+1 facing, 2: 2+2 facing.

DTSO Lot No. 30955 York 1981–82. –/52 + catering point. 31.4 t.
MSO. Lot No. 30958 York 1981–82. –/62. 51.3 t.
TSO. Lot No. 30957 Derby 1981–82. –/42(5) 1W 1T 1TD. 30.2 t.
DTCO Lot No. 30956 York 1981–82. 22/16 + catering point. 31.6 t.

317708	**NX**	A		ZG	77007	62668	71584	77055
317709	**NX**	A		ZG	77008	62669	71585	77056
317710	**NX**	A		ZG	77009	62670	71586	77057
317714	**NX**	A		ZG	77013	62674	71590	77061
317719	**NX**	A		ZG	77018	62679	71595	77066
317722	**NX**	A		IL	77021	62682	71598	77069
317723	**NX**	A		ZG	77022	62683	71599	77070
317729	**NX**	A		ZG	77028	62689	71605	77076
317732	**NX**	A		ZG	77031	62692	71608	77079

Names (carried on DTCO):

317709 Len Camp | 317723 The Tottenham Flyer

Class 317/8. Pressure Ventilated. Units refurbished and renumbered from Class 317/1 in 2005–06 at Wabtec, Doncaster for use on Stansted Express services. Fitted with luggage stacks. Displaced from Stansted services in 2011 by Class 379s.

Formation: DTSO–MSO–TCO–DTSO.
Seating Layout: 1: 2+2 facing, 2: 3+2 facing.

DTSO(A) Lot No. 30955 York 1981–82. –/66. 29.5 t.
MSO. Lot No. 30958 York 1981–82. –/71. 49.0 t.
TCO. Lot No. 30957 Derby 1981–82. 20/42 2T. 29.0 t.
DTSO(B) Lot No. 30956 York 1981–82. –/66. 29.5 t.

317881	NX	A	*EA*	IL	77020	62681	71597	77068	
317882	NC	A	*EA*	IL	77023	62684	71600	77071	
317883	NC	A	*EA*	IL	77000	62685	71601	77072	
317884	NC	A	*EA*	IL	77025	62686	71602	77073	
317885	NC	A	*EA*	IL	77026	62687	71603	77074	
317886	NC	A	*EA*	IL	77027	62688	71604	77075	
317887	NX	A	*EA*	IL	77043	62704	71606	77077	
317888	NX	A	*EA*	IL	77030	62691	71607	77078	
317889	NX	A	*EA*	IL	77032	62693	71609	77080	
317890	NX	A	*EA*	IL	77033	62694	71610	77081	
317891	NX	A	*EA*	IL	77034	62695	71611	77082	
317892	NX	A	*EA*	IL	77035	62696	71612	77083	Ilford Depot

CLASS 318   BREL YORK

Outer suburban units.

Formation: DTSO–MSO–DTSO.
Construction: Steel.
Traction Motors: Four Brush TM 2141 of 268 kW.
Wheel Arrangement: 2-2 + Bo-Bo + 2-2.
Braking: Disc. **Dimensions:** 19.83/19.92 x 2.82 m.
Bogies: BP20 (MSO), BT13 (others). **Couplers:** Tightlock.
Gangways: Within unit. **Control System:** Thyristor.
Doors: Sliding. **Maximum Speed:** 90 mph.
Seating Layout: 3+2 facing.
Multiple Working: Within class & with Classes 317, 319, 320, 321, 322 and 323.

77240–259. DTSO. Lot No. 30999 1985–86. –/64 1T. 30.0 t.
77288. DTSO. Lot No. 31020 1987. –/64 1T. 30.0 t.
62866–885. MSO. Lot No. 30998 1985–86. –/77. 50.9 t.
62890. MSO. Lot No. 31019 1987. –/77. 50.9 t.
77260–279. DTSO. Lot No. 31000 1985–86. –/72. 29.6 t.
77289. DTSO. Lot No. 31021 1987. –/72. 29.6 t.

318250	SC	E	*SR*	GW	77240	62866	77260
318251	SC	E	*SR*	GW	77241	62867	77261
318252	SC	E	*SR*	GW	77242	62868	77262
318253	SC	E	*SR*	GW	77243	62869	77263
318254	SC	E	*SR*	GW	77244	62870	77264
318255	SC	E	*SR*	GW	77245	62871	77265

▲ First Capital Connect's 313s are used on local Great Northern services into Moorgate and King's Cross. On 13/05/12 313 052 pauses at Hadley Wood working the 15.28 (Sun) Welwyn Garden City–King's Cross. **Robert Pritchard**

▼ Carrying the new Greater Anglia white livery with red doors, 317 661 brings up the rear of a diverted 16.04 Cambridge–Liverpool Street at Lea Bridge on 12/05/12 (the train was led by 317 881). **Antony Guppy**

▲ At the time of writing all of the Class 318s still carried Strathclyde PTE carmine & cream livery. On 08/08/12 318 254 leads a Class 320 on the Lanark branch with the 17.04 Lanark–Anderson. **Robin Ralston**

▼ First Capital Connect-liveried 319 437 and 319 451 approach Hendon with the 16.16 Elephant & Castle–Bedford on 24/07/12. **Antony Guppy**

▲ ScotRail Saltire-liveried 320 314 has just left Carluke station working the 14.23 Dalmuir–Lanark on 10/08/12. **Robin Ralston**

▼ Greater Anglia-liveried 321 364 leads a 12-car Ipswich–Colchester empty stock working also formed of 321 305 and 321 443 on the approach to Manningtree on 12/07/12. **Robert Pritchard**

▲ London Midland-liveried 323 204 arrives at Aston working the 14.01 Walsall–Wolverhampton on 24/07/12. **Cliff Beeton**

▼ A trio of Royal Mail Class 325 sets, led by 325 002, pass Crewe working the 1S96 16.40 Willesden–Shieldmuir postal on 16/05/12. **Cliff Beeton**

▲ Refurbished Heathrow Express unit 332 011 stands outside Old Oak Common depot on 24/07/12. **Brian Denton**

▼ In the joint West Yorkshire PTE/Northern livery, 333 016 pauses at Burley-in-Wharfedale with the 13.51 Ilkley–Bradford Forster Square on 22/04/11.
Robert Pritchard

▲ ScotRail-liveried 334 020 passes Hillend Loch, near Caldercruix on the Airdrie–Bathgate line, with the 14.37 Edinburgh–Milngavie on 03/09/12. **Ian Lothian**

▼ London Midland-liveried 350 257 arrives at Rugby with the 12.33 Birmingham New Street–London Euston on 26/07/12. **Stuart Armstrong**

▲ National Express-liveried 357 216 and 357 204 arrive at Leigh-on-Sea with the 11.00 London Fenchurch Street–Shoeburyness on 09/08/12. **William Turvill**

▼ Heathrow Connect 360 204 passes Friars Junction, Acton, working the 12.03 London Paddington–Heathrow Airport on 31/08/12. **Antony Guppy**

▲ Still carrying the Connex inspired white and black livery with yellow doors, 375 803 and 375 918 pass Petts Wood with the 09.08 London Charing Cross–Canterbury West/Ramsgate on 01/07/12. **Robert Pritchard**

▼ In the same livery, 376 025 and 375 029 approach Albany Park with the 11.10 Cannon Street–Cannon Street via Slade Green circular 11/08/12. **William Turvill**

318256	**SC**	E	*SR*	GW	77246	62872	77266	
318257	**SC**	E	*SR*	GW	77247	62873	77267	
318258	**SC**	E	*SR*	GW	77248	62874	77268	
318259	**SC**	E	*SR*	GW	77249	62875	77269	Citizens' Network
318260	**SC**	E	*SR*	GW	77250	62876	77270	
318261	**SC**	E	*SR*	GW	77251	62877	77271	
318262	**SC**	E	*SR*	GW	77252	62878	77272	
318263	**SC**	E	*SR*	GW	77253	62879	77273	
318264	**SC**	E	*SR*	GW	77254	62880	77274	
318265	**SC**	E	*SR*	GW	77255	62881	77275	
318266	**SC**	E	*SR*	GW	77256	62882	77276	STRATHCLYDER
318267	**SC**	E	*SR*	GW	77257	62883	77277	
318268	**SC**	E	*SR*	GW	77258	62884	77278	
318269	**SC**	E	*SR*	GW	77259	62885	77279	
318270	**SC**	E	*SR*	GW	77288	62890	77289	

CLASS 319   BREL YORK

Express and outer suburban units.

Formation: Various, see sub-class headings.
Systems: 25 kV AC overhead/750 V DC third rail.
Construction: Steel.
Traction Motors: Four GEC G315BZ of 268 kW.
Wheel Arrangement: 2-2 + Bo-Bo + 2-2 + 2-2.
Braking: Disc.   **Dimensions:** 20.17/20.16 x 2.82 m.
Bogies: P7-4 (MSO), T3-7 (others).   **Couplers:** Tightlock.
Gangways: Within unit + end doors.   **Control System:** GTO chopper.
Doors: Sliding.   **Maximum Speed:** 100 mph.
Seating Layout: Various, see sub-class headings.
Multiple Working: Within class & with Classes 317, 318, 320, 321, 322 and 323.

Class 319/0. DTSO–MSO–TSO–DTSO.

Seating Layout: 3+2 facing.

DTSO(A). Lot No. 31022 (odd nos.) 1987–88. –/82. 28.2 t.
MSO. Lot No. 31023 1987–88. –/82. 49.2 t.
TSO. Lot No. 31024 1987–88. –/77 2T. 31.0 t.
DTSO(B). Lot No. 31025 (even nos.) 1987–88. –/78. 28.1 t.

319001	**FU**	P	*FC*	BF	77291	62891	71772	77290
319002	**FU**	P	*FC*	BF	77293	62892	71773	77292
319003	**FU**	P	*FC*	BF	77295	62893	71774	77294
319004	**FU**	P	*FC*	BF	77297	62894	71775	77296
319005	**FU**	P	*FC*	BF	77299	62895	71776	77298
319006	**FU**	P	*FC*	BF	77301	62896	71777	77300
319007	**FU**	P	*FC*	BF	77303	62897	71778	77302
319008	**SN**	P	*FC*	BF	77305	62898	71779	77304
319009	**SN**	P	*FC*	BF	77307	62899	71780	77306
319010	**FU**	P	*FC*	BF	77309	62900	71781	77308
319011	**SN**	P	*FC*	BF	77311	62901	71782	77310
319012	**SN**	P	*FC*	BF	77313	62902	71783	77312
319013	**SN**	P	*FC*	BF	77315	62903	71784	77314

Names (carried on TSO):

319 008 Cheriton	319 011 John Ruskin College
319 009 Coquelles	319 013 The Surrey Hills

Class 319/2. DTSO–MSO–TSO–DTCO. Units converted from Class 319/0.

Seating Layout: 1: 2+1 facing, 2: 2+2 facing.

DTSO. Lot No. 31022 (odd nos.) 1987–88. –/64. 28.2 t.
MSO. Lot No. 31023 1987–88. –/73 2T. 49.2 t.
TSO. Lot No. 31024 1987–88. –/52 1T 1TD. 31.0 t.
DTCO. Lot No. 31025 (even nos.) 1987–88. 18/36. 28.1 t.

Advertising liveries: 319 215 Visit Switzerland (red).
319 218 Lycamobile (white).

319215	**SN**	P	*FC*	BF	77317	62904	71785	77316
319215	**AL**	P	*FC*	BF	77319	62905	71786	77318
319216	**SN**	P	*FC*	BF	77321	62906	71787	77320
319217	**SN**	P	*FC*	BF	77323	62907	71788	77322 Brighton
319218	**AL**	P	*FC*	BF	77325	62908	71789	77324 Croydon
319219	**SN**	P	*FC*	BF	77327	62909	71790	77326
319220	**SN**	P	*FC*	BF	77329	62910	71791	77328

Class 319/3. DTSO–MSO–TSO–DTSO. Converted from Class 319/1 by replacing First Class seats with Standard Class seats. Used mainly on the Luton–Sutton/Wimbledon routes.

Seating Layout: 3+2 facing.
Dimensions: 19.33 x 2.82 m.

Advertising livery: 319 364 & 319 365 Thameslink Programme (multi-coloured horizontal stripes with pink ends).

DTSO(A). Lot No. 31063 1990. –/70. 29.0 t.
MSO. Lot No. 31064 1990. –/78. 50.6 t.
TSO. Lot No. 31065 1990. –/74 2T. 31.0 t.
DTSO(B). Lot No. 31066 1990. –/75. 29.7 t.

319361	**FU**	P	*FC*	BF	77459	63043	71929	77458
319362	**FU**	P	*FC*	BF	77461	63044	71930	77460
319363	**FU**	P	*FC*	BF	77463	63045	71931	77462
319364	**AL**	P	*FC*	BF	77465	63046	71932	77464
319365	**AL**	P	*FC*	BF	77467	63047	71933	77466
319366	**FU**	P	*FC*	BF	77469	63048	71934	77468
319367	**FU**	P	*FC*	BF	77471	63049	71935	77470
319368	**FU**	P	*FC*	BF	77473	63050	71936	77472
319369	**FU**	P	*FC*	BF	77475	63051	71937	77474
319370	**FU**	P	*FC*	BF	77477	63052	71938	77476
319371	**FU**	P	*FC*	BF	77479	63053	71939	77478
319372	**FU**	P	*FC*	BF	77481	63054	71940	77480
319373	**FU**	P	*FC*	BF	77483	63055	71941	77482
319374	**FU**	P	*FC*	BF	77485	63056	71942	77484
319375	**FU**	P	*FC*	BF	77487	63057	71943	77486
319376	**FU**	P	*FC*	BF	77489	63058	71944	77488

319377	**FU**	P	*FC*	BF	77491	63059	71945	77490
319378	**FU**	P	*FC*	BF	77493	63060	71946	77492
319379	**FU**	P	*FC*	BF	77495	63061	71947	77494
319380	**FU**	P	*FC*	BF	77497	63062	71948	77496
319381	**FU**	P	*FC*	BF	77973	63093	71979	77974
319382	**FU**	P	*FC*	BF	77975	63094	71980	77976
319383	**FU**	P	*FC*	BF	77977	63095	71981	77978
319384	**FU**	P	*FC*	BF	77979	63096	71982	77980
319385	**FU**	P	*FC*	BF	77981	63097	71983	77982
319386	**FU**	P	*FC*	BF	77983	63098	71984	77984

Names (carried on TSO):

319364 Transforming Blackfriars
319365 Transforming Farringdon
319374 Bedford Cauldwell TMD

Class 319/4. DTCO–MSO–TSO–DTSO. Converted from Class 319/0. Refurbished with carpets. DTSO(A) converted to composite. Used mainly on the Bedford–Gatwick–Brighton route.

Seating Layout: 1: 2+1 facing 2: 2+2/3+2 facing.

77331–381. DTCO. Lot No. 31022 (odd nos.) 1987–88. 12/51. 28.2t.
77431–457. DTCO. Lot No. 31038 (odd nos.) 1988. 12/51. 28.2t.
62911–936. MSO. Lot No. 31023 1987–88. –/74. 49.2t.
62961–974. MSO. Lot No. 31039 1988. –/74. 49.2t.
71792–817. TSO. Lot No. 31024 1987–88. –/67 2T. 31.0t.
71866–879. TSO. Lot No. 31040 1988. –67 2T. 31.0t.
77330–380. DTSO. Lot No. 31025 (even nos.) 1987–88. –/71 1W. 28.1t.
77430–456. DTSO. Lot No. 31041 (even nos.) 1988. –/71 1W. 28.1t.

319421	**FU**	P	*FC*	BF	77331	62911	71792	77330
319422	**FU**	P	*FC*	BF	77333	62912	71793	77332
319423	**FU**	P	*FC*	BF	77335	62913	71794	77334
319424	**FU**	P	*FC*	BF	77337	62914	71795	77336
319425	**FU**	P	*FC*	BF	77339	62915	71796	77338
319426	**FU**	P	*FC*	BF	77341	62916	71797	77340
319427	**FU**	P	*FC*	BF	77343	62917	71798	77342
319428	**FU**	P	*FC*	BF	77345	62918	71799	77344
319429	**FU**	P	*FC*	BF	77347	62919	71800	77346
319430	**FU**	P	*FC*	BF	77349	62920	71801	77348
319431	**FU**	P	*FC*	BF	77351	62921	71802	77350
319432	**FU**	P	*FC*	BF	77353	62922	71803	77352
319433	**FU**	P	*FC*	BF	77355	62923	71804	77354
319434	**FU**	P	*FC*	BF	77357	62924	71805	77356
319435	**FU**	P	*FC*	BF	77359	62925	71806	77358
319436	**FU**	P	*FC*	BF	77361	62926	71807	77360
319437	**FU**	P	*FC*	BF	77363	62927	71808	77362
319438	**FU**	P	*FC*	BF	77365	62928	71809	77364
319439	**FU**	P	*FC*	BF	77367	62929	71810	77366
319440	**FU**	P	*FC*	BF	77369	62930	71811	77368
319441	**FU**	P	*FC*	BF	77371	62931	71812	77370
319442	**FU**	P	*FC*	BF	77373	62932	71813	77372

319443	**FU**	P	*FC*	BF	77375	62933	71814	77374
319444	**FU**	P	*FC*	BF	77377	62934	71815	77376
319445	**FU**	P	*FC*	BF	77379	62935	71816	77378
319446	**FU**	P	*FC*	BF	77381	62936	71817	77380
319447	**FU**	P	*FC*	BF	77431	62961	71866	77430
319448	**FU**	P	*FC*	BF	77433	62962	71867	77432
319449	**FU**	P	*FC*	BF	77435	62963	71868	77434
319450	**FU**	P	*FC*	BF	77437	62964	71869	77436
319451	**FU**	P	*FC*	BF	77439	62965	71870	77438
319452	**FU**	P	*FC*	BF	77441	62966	71871	77440
319453	**FU**	P	*FC*	BF	77443	62967	71872	77442
319454	**FU**	P	*FC*	BF	77445	62968	71873	77444
319455	**FU**	P	*FC*	BF	77447	62969	71874	77446
319456	**FU**	P	*FC*	BF	77449	62970	71875	77448
319457	**FU**	P	*FC*	RF	77451	62971	71876	77450
319458	**FU**	P	*FC*	BF	77453	62972	71877	77452
319459	**FU**	P	*FC*	BF	77455	62973	71878	77454
319460	**FU**	P	*FC*	BF	77457	62974	71879	77456

Names (carried on TSO):

319425	Transforming Travel
319435	Adrian Jackson-Robbins Chairman 1987–2007 Association of Public Transport Users
319441	Driver Mick Winnett
319446	St Pancras International
319448	Elstree Studios The Home of British Film and Television production
319449	King's Cross Thameslink

CLASS 320   BREL YORK

Suburban units.

Formation: DTSO–MSO–DTSO.
Construction: Steel
Traction Motors: Four Brush TM2141B of 268 kW.
Wheel Arrangement: 2-2 + Bo-Bo + 2-2.
Braking: Disc.   **Dimensions:** 19.95 x 2.82 m.
Bogies: P7-4 (MSO), T3-7 (others).   **Couplers:** Tightlock.
Gangways: Within unit.   **Control System:** Thyristor.
Doors: Sliding.   **Maximum Speed:** 90 mph.
Seating Layout: 3+2 facing.
Multiple Working: Within class & with Classes 317, 318, 319, 321, 322 and 323.

DTSO (A). Lot No. 31060 1990. –/76 1W (* –/51(4) 1TD 2W). 29.1 t (* 41.0 t).
MSO. Lot No. 31062 1990. –/76 1W (* –/78). 52.1 t.
DTSO (B). Lot No. 31061 1990. –/75 (* –/77). 31.7 t.

Note: Units marked * have been fitted with a toilet in the DTSO (A) vehicle.

320301	*	**SR**	E	*SR*	GW	77899	63021	77921
320302	*	**SR**	E	*SR*	GW	77900	63022	77922
320303	*	**SR**	E	*SR*	GW	77901	63023	77923
320304	*	**SR**	E	*SR*	GW	77902	63024	77924

320305		SC	E	SR	GW	77903	63025	77925
320306	*	SR	E	SR	GW	77904	63026	77926
320307		SC	E	SR	GW	77905	63027	77927
320308	*	SR	E	SR	GW	77906	63028	77928
320309		SC	E	SR	GW	77907	63029	77929
320310		SC	E	SR	GW	77908	63030	77930
320311	*	SR	E	SR	GW	77909	63031	77931
320312		SC	E	SR	GW	77910	63032	77932
320313	*	SR	E	SR	GW	77911	63033	77933
320314	*	SR	E	SR	GW	77912	63034	77934
320315	*	SR	E	SR	GW	77913	63035	77935
320316	*	SR	E	SR	GW	77914	63036	77936
320317		SC	E	SR	GW	77915	63037	77937
320318	*	SR	E	SR	GW	77916	63038	77938
320319		SC	E	SR	GW	77917	63039	77939
320320		SC	E	SR	GW	77918	63040	77940
320321	*	SR	E	SR	GW	77919	63041	77941
320322		SC	E	SR	GW	77920	63042	77942

Names (carried on MSO, all to be removed on repainting):

320305 GLASGOW SCHOOL OF ART 1845 150 1995
320309 Radio Clyde 25th Anniversary
320312 Sir William A Smith Founder of the Boys' Brigade
320322 Festive Glasgow Orchid

CLASS 321   BREL YORK

Outer suburban units.

Formation: DTCO (DTSO on Class 321/9)–MSO–TSO–DTSO.
Construction: Steel.
Traction Motors: Four Brush TM2141C (268 kW).
Wheel Arrangement: 2-2 + Bo-Bo + 2-2 + 2-2.
Braking: Disc.   **Dimensions:** 19.95 x 2.82 m.
Bogies: P7-4 (MSO), T3-7 (others).   **Couplers:** Tightlock.
Gangways: Within unit.   **Control System:** Thyristor.
Doors: Sliding.   **Maximum Speed:** 100 mph.
Seating Layout: 1: 2+2 facing, 2: 3+2 facing.
Multiple Working: Within class & with Classes 317, 318, 319, 320, 322 and 323.

Class 321/3.

DTCO. Lot No. 31053 1988–90. 16/57. 29.7 t.
MSO. Lot No. 31054 1988–90. –/82. 51.5 t.
TSO. Lot No. 31055 1988–90. –/75 2T. 29.1 t.
DTSO. Lot No. 31056 1988–90. –/78. 29.7 t.

321301	NX	E	EA	IL	78049	62975	71880	77853
321302	NX	E	EA	IL	78050	62976	71881	77854
321303	NX	E	EA	IL	78051	62977	71882	77855
321304	NX	E	EA	IL	78052	62978	71883	77856
321305	NX	E	EA	IL	78053	62979	71884	77857
321306	NX	E	EA	IL	78054	62980	71885	77858

321 307	**NX**	E	*EA*	IL	78055	62981	71886	77859
321 308	**NX**	E	*EA*	IL	78056	62982	71887	77860
321 309	**NX**	E	*EA*	IL	78057	62983	71888	77861
321 310	**NX**	E	*EA*	IL	78058	62984	71889	77862
321 311	**NX**	E	*EA*	IL	78059	62985	71890	77863
321 312	**NX**	E	*EA*	IL	78060	62986	71891	77864
321 313	**NX**	E	*EA*	IL	78061	62987	71892	77865
321 314	**NX**	E	*EA*	IL	78062	62988	71893	77866
321 315	**NX**	E	*EA*	IL	78063	62989	71894	77867
321 316	**NX**	E	*EA*	IL	78064	62990	71895	77868
321 317	**NX**	E	*EA*	IL	78065	62991	71896	77869
321 318	**NX**	E	*EA*	IL	78066	62992	71897	77870
321 319	**NX**	E	*EA*	IL	78067	62993	71898	77871
321 320	**NX**	E	*EA*	IL	78068	62994	71899	77872
321 321	**NX**	E	*EA*	IL	78069	62995	71900	77873
321 322	**NX**	E	*EA*	IL	78070	62996	71901	77874
321 323	**NX**	E	*EA*	IL	78071	62997	71902	77875
321 324	**NX**	E	*EA*	IL	78072	62998	71903	77876
321 325	**NX**	E	*EA*	IL	78073	62999	71904	77877
321 326	**NX**	E	*EA*	IL	78074	63000	71905	77878
321 327	**NC**	E	*EA*	IL	78075	63001	71906	77879
321 328	**NX**	E	*EA*	IL	78076	63002	71907	77880
321 329	**NX**	E	*EA*	IL	78077	63003	71908	77881
321 330	**NC**	E	*EA*	IL	78078	63004	71909	77882
321 331	**NC**	E	*EA*	IL	78079	63005	71910	77883
321 332	**NC**	E	*EA*	IL	78080	63006	71911	77884
321 333	**NC**	E	*EA*	IL	78081	63007	71912	77885
321 334	**NC**	E	*EA*	IL	78082	63008	71913	77886
321 335	**NC**	E	*EA*	IL	78083	63009	71914	77887
321 336	**NC**	E	*EA*	IL	78084	63010	71915	77888
321 337	**NC**	E	*EA*	IL	78085	63011	71916	77889
321 338	**NC**	E	*EA*	IL	78086	63012	71917	77890
321 339	**NC**	E	*EA*	IL	78087	63013	71918	77891
321 340	**NC**	E	*EA*	IL	78088	63014	71919	77892
321 341	**NC**	E	*EA*	IL	78089	63015	71920	77893
321 342	**NC**	E	*EA*	IL	78090	63016	71921	77894
321 343	**NC**	E	*EA*	IL	78091	63017	71922	77895
321 344	**NC**	E	*EA*	IL	78092	63018	71923	77896
321 345	**NC**	E	*EA*	IL	78093	63019	71924	77897
321 346	**NC**	E	*EA*	IL	78094	63020	71925	77898
321 347	**NC**	E	*EA*	IL	78131	63105	71991	78280
321 348	**NC**	E	*EA*	IL	78132	63106	71992	78281
321 349	**NC**	E	*EA*	IL	78133	63107	71993	78282
321 350	**NC**	E	*EA*	IL	78134	63108	71994	78283
321 351	**NC**	E	*EA*	IL	78135	63109	71995	78284
321 352	**NC**	E	*EA*	IL	78136	63110	71996	78285
321 353	**NC**	E	*EA*	IL	78137	63111	71997	78286
321 354	**NC**	E	*EA*	IL	78138	63112	71998	78287
321 355	**NC**	E	*EA*	IL	78139	63113	71999	78288
321 356	**NC**	E	*EA*	IL	78140	63114	72000	78289
321 357	**NC**	E	*EA*	IL	78141	63115	72001	78290

321358	**NC**	E	*EA*	IL	78142	63116	72002	78291
321359	**GA**	E	*EA*	IL	78143	63117	72003	78292
321360	**NC**	E	*EA*	IL	78144	63118	72004	78293
321361	**GA**	E	*EA*	IL	78145	63119	72005	78294
321362	**GA**	E	*EA*	IL	78146	63120	72006	78295
321363	**GA**	E	*EA*	IL	78147	63121	72007	78296
321364	**GA**	E	*EA*	IL	78148	63122	72008	78297
321365	**GA**	E	*EA*	IL	78149	63123	72009	78298
321366	**GA**	E	*EA*	IL	78150	63124	72010	78299

Names (carried on TSO):

321312 Southend-on-Sea
321313 University of Essex
321321 NSPCC ESSEX FULL STOP
321334 Amsterdam
321336 GEOFFREY FREEMAN ALLEN
321342 R. Barnes
321343 RSA RAILWAY STUDY ASSOCIATION
321351 London Southend Airport
321361 Phoenix

Class 321/4.

DTCO. Lot No. 31067 1989–90. 28/40. 29.8 t.
MSO. Lot No. 31068 1989–90. –/79. 51.6 t.
TSO. Lot No. 31069 1989–90. –/74 2T. 29.2 t.
DTSO. Lot No. 31070 1989–90. –/78. 29.8 t.

Notes: The original vehicles 71966 and 77960 from 321418 and 78114 and 63082 from 321420 were written off after the Watford Junction accident in 1996. The undamaged vehicles were formed together as 321418 whilst four new vehices were built in 1997, taking the same numbers as the scrapped vehicles, and these became the second 321420.

The DTCOs of 321438–448 have had 12 First Class seats declassified.

321401	**FU**	E	*FC*	HE	78095	63063	71949	77943
321402	**FU**	E	*FC*	HE	78096	63064	71950	77944
321403	**FU**	E	*FC*	HE	78097	63065	71951	77945
321404	**FU**	E	*FC*	HE	78098	63066	71952	77946
321405	**FU**	E	*FC*	HE	78099	63067	71953	77947
321406	**FU**	E	*FC*	HE	78100	63068	71954	77948
321407	**FU**	E	*FC*	HE	78101	63069	71955	77949
321408	**FU**	E	*FC*	HE	78102	63070	71956	77950
321409	**FU**	E	*FC*	HE	78103	63071	71957	77951
321410	**FU**	E	*FC*	HE	78104	63072	71958	77952
321411	**LM**	E	*LM*	NN	78105	63073	71959	77953
321412	**LM**	E	*LM*	NN	78106	63074	71960	77954
321413	**LM**	E	*LM*	NN	78107	63075	71961	77955
321414	**LM**	E	*LM*	NN	78108	63076	71962	77956
321415	**LM**	E	*LM*	NN	78109	63077	71963	77957
321416	**LM**	E	*LM*	NN	78110	63078	71964	77958
321417	**LM**	E	*LM*	NN	78111	63079	71965	77959

321418	**FU**	E	*FC*	HE	78112	63080	71968	77962
321419	**FU**	E	*FC*	HE	78113	63081	71967	77961
321420	**FU**	E	*FC*	HE	78114	63082	71966	77960
321421	**NC**	E	*EA*	IL	78115	63083	71969	77963
321422	**NC**	E	*EA*	IL	78116	63084	71970	77964
321423	**NC**	E	*EA*	IL	78117	63085	71971	77965
321424	**NX**	E	*EA*	IL	78118	63086	71972	77966
321425	**NC**	E	*EA*	IL	78119	63087	71973	77967
321426	**NX**	E	*EA*	IL	78120	63088	71974	77968
321427	**NX**	E	*EA*	IL	78121	63089	71975	77969
321428	**NX**	E	*EA*	IL	78122	63090	71976	77970
321429	**NX**	E	*EA*	IL	78123	63091	71977	77971
321430	**NX**	E	*EA*	IL	78124	63092	71978	77972
321431	**NX**	E	*EA*	IL	78151	63125	72011	78300
321432	**NC**	E	*EA*	IL	78152	63126	72012	78301
321433	**NC**	E	*EA*	IL	78153	63127	72013	78302
321434	**NC**	E	*EA*	IL	78154	63128	72014	78303
321435	**NC**	E	*EA*	IL	78155	63129	72015	78304
321436	**NC**	E	*EA*	IL	78156	63130	72016	78305
321437	**NC**	E	*EA*	IL	78157	63131	72017	78306
321438	**GA**	E	*EA*	IL	78158	63132	72018	78307
321439	**GA**	E	*EA*	IL	78159	63133	72019	78308
321440	**GA**	E	*EA*	IL	78160	63134	72020	78309
321441	**GA**	E	*EA*	IL	78161	63135	72021	78310
321442	**GA**	E	*EA*	IL	78162	63136	72022	78311
321443	**GA**	E	*EA*	IL	78125	63099	71985	78274
321444	**NC**	E	*EA*	IL	78126	63100	71986	78275
321445	**NC**	E	*EA*	IL	78127	63101	71987	78276
321446	**NC**	E	*EA*	IL	78128	63102	71988	78277
321447	**GA**	E	*EA*	IL	78129	63103	71989	78278
321448	**GA**	E	*EA*	IL	78130	63104	71990	78279

Names (carried on TSO):

321403 Stewart Fleming Signalman King's Cross
321428 The Essex Commuter
321444 Essex Lifeboats
321446 George Mullings

Class 321/9. DTSO(A)–MSO–TSO–DTSO(B).

DTSO(A). Lot No. 31108 1991. –/70(8). 29.2 t.
MSO. Lot No. 31109 1991. –/79. 51.1 t.
TSO. Lot No. 31110 1991. –/74 2T. 29.0 t.
DTSO(B). Lot No. 31111 1991. –/70(7) 1W. 29.2 t.

321901	**YR**	E	*NO*	NL	77990	63153	72128	77993
321902	**YR**	E	*NO*	NL	77991	63154	72129	77994
321903	**YR**	E	*NO*	NL	77992	63155	72130	77995

CLASS 322   BREL YORK

Units built for use on Stansted Airport services, used for a number of years with ScotRail before transfer to Northern.

Formation: DTSO–MSO–TSO–DTSO.
Construction: Steel.
Traction Motors: Four Brush TM2141C (268 kW).
Wheel Arrangement: 2-2 + Bo-Bo + 2-2 + 2-2.
Braking: Disc.   **Dimensions:** 19.95/19.92 x 2.82 m.
Bogies: P7-4 (MSO), T3-7 (others).   **Couplers:** Tightlock.
Gangways: Within unit.   **Control System:** Thyristor.
Doors: Sliding.   **Maximum Speed:** 100 mph.
Seating Layout: 3+2 facing.
Multiple Working: Within class & with Classes 317, 318, 319, 320, 321 and 323.

DTSO(A). Lot No. 31094 1990. –/58. 29.3 t.
MSO. Lot No. 31092 1990. –/83. 51.5 t.
TSO. Lot No. 31093 1990. –/76 2T. 28.8 t.
DTSO(B). Lot No. 31091 1990. –/89(3) 1W. 29.1 t.

322 481	**FB**	E	*NO*	NL	78163	63137	72023	77985
322 482	**FB**	E	*NO*	NL	78164	63138	72024	77986
322 483	**FB**	E	*NO*	NL	78165	63139	72025	77987
322 484	**FB**	E	*NO*	NL	78166	63140	72026	77988
322 485	**FB**	E	*NO*	NL	78167	63141	72027	77989

CLASS 323 HUNSLET TRANSPORTATION PROJECTS

Suburban units.

Formation: DMSO–PTSO–DMSO.
Construction: Welded aluminium alloy.
Traction Motors: Four Holec DMKT 52/24 asynchronous of 146 kW.
Wheel Arrangement: Bo-Bo + 2-2 + Bo-Bo.
Braking: Disc.   **Dimensions:** 23.37/23.44 x 2.80 m.
Bogies: SRP BP62 (DMSO), BT52 (PTSO). **Couplers:** Tightlock.
Gangways: Within unit.   **Control System:** GTO Inverter.
Doors: Sliding plug.   **Maximum Speed:** 90 mph.
Seating Layout: 3+2 facing/unidirectional.
Multiple Working: Within class & with Classes 317, 318, 319, 320, 321 and 322.

DMSO(A). Lot No. 31112 Hunslet 1992–93. –/98 (* –/82). 41.0 t.
TSO. Lot No. 31113 Hunslet 1992–93. –/88(5) 1T 2W. (* –/80 1T 2W). 39.4 t.
DMSO(B). Lot No. 31114 Hunslet 1992–93. –/98 (* –/82). 41.0 t.

323 201	**LM**	P	*LM*	SO	64001	72201	65001
323 202	**LM**	P	*LM*	SO	64002	72202	65002
323 203	**LM**	P	*LM*	SO	64003	72203	65003
323 204	**LM**	P	*LM*	SO	64004	72204	65004
323 205	**LM**	P	*LM*	SO	64005	72205	65005
323 206	**LM**	P	*LM*	SO	64006	72206	65006
323 207	**LM**	P	*LM*	SO	64007	72207	65007
323 208	**LM**	P	*LM*	SO	64008	72208	65008

323 209	**LM**	P	*LM*	SO	64009	72209	65009
323 210	**LM**	P	*LM*	SO	64010	72210	65010
323 211	**LM**	P	*LM*	SO	64011	72211	65011
323 212	**LM**	P	*LM*	SO	64012	72212	65012
323 213	**LM**	P	*LM*	SO	64013	72213	65013
323 214	**LM**	P	*LM*	SO	64014	72214	65014
323 215	**LM**	P	*LM*	SO	64015	72215	65015
323 216	**LM**	P	*LM*	SO	64016	72216	65016
323 217	**LM**	P	*LM*	SO	64017	72217	65017
323 218	**LM**	P	*LM*	SO	64018	72218	65018
323 219	**LM**	P	*LM*	SO	64019	72219	65019
323 220	**LM**	P	*LM*	SO	64020	72220	65020
323 221	**LM**	P	*LM*	SO	64021	72221	65021
323 222	**LM**	P	*LM*	SO	64022	72222	65022
323 223	*	**NO**	P	*NO*	LG	64023	72223	65023
323 224	*	**NO**	P	*NO*	LG	64024	72224	65024
323 225	*	**NO**	P	*NO*	LG	64025	72225	65025
323 226	**NO**	P	*NO*	LG	64026	72226	65026
323 227	**NO**	P	*NO*	LG	64027	72227	65027
323 228	**NO**	P	*NO*	LG	64028	72228	65028
323 229	**NO**	P	*NO*	LG	64029	72229	65029
323 230	**NO**	P	*NO*	LG	64030	72230	65030
323 231	**NO**	P	*NO*	LG	64031	72231	65031
323 232	**NO**	P	*NO*	LG	64032	72232	65032
323 233	**NO**	P	*NO*	LG	64033	72233	65033
323 234	**NO**	P	*NO*	LG	64034	72234	65034
323 235	**NO**	P	*NO*	LG	64035	72235	65035
323 236	**NO**	P	*NO*	LG	64036	72236	65036
323 237	**NO**	P	*NO*	LG	64037	72237	65037
323 238	**NO**	P	*NO*	LG	64038	72238	65038
323 239	**NO**	P	*NO*	LG	64039	72239	65039
323 240	**LM**	P	*LM*	SO	64040	72340	65040
323 241	**LM**	P	*LM*	SO	64041	72341	65041
323 242	**LM**	P	*LM*	SO	64042	72342	65042
323 243	**LM**	P	*LM*	SO	64043	72343	65043

CLASS 325   ABB DERBY

Postal units based on Class 319s. Compatible with diesel or electric locomotive haulage.

Formation: DTPMV–MPMV–TPMV–DTPMV.
System: 25 kV AC overhead/750 V DC third rail.
Construction: Steel.
Traction Motors: Four GEC G315BZ of 268 kW.
Wheel Arrangement: 2-2 + Bo-Bo + 2-2 + 2-2.
Braking: Disc.   **Dimensions:** 19.33 x 2.82 m.
Bogies: P7-4 (MSO), T3-7 (others).   **Couplers:** Drop-head buckeye.
Gangways: None.   **Control System:** GTO Chopper.
Doors: Roller shutter.   **Maximum Speed:** 100 mph.
Multiple Working: Within class.

DTPMV. Lot No. 31144 1995. 29.1 t.
MPMV. Lot No. 31145 1995. 49.5 t.
TPMV. Lot No. 31146 1995. 30.7 t.

325 001	**RM**	RM	*DB*	CE	68300	68340	68360	68301
325 002	**RM**	RM	*DB*	CE	68302	68341	68361	68303
325 003	**RM**	RM	*DB*	CE	68304	68342	68362	68305
325 004	**RM**	RM	*DB*	CE	68306	68343	68363	68307
325 005	**RM**	RM	*DB*	CE	68308	68344	68364	68309
325 006	**RM**	RM	*DB*	CE	68310	68345	68365	68311
325 007	**RM**	RM	*DB*	CE	68312	68346	68366	68313
325 008	**RM**	RM	*DB*	CE	68314	68347	68367	68315
325 009	**RM**	RM	*DB*	CE	68316	68349	68368	68317
325 011	**RM**	RM	*DB*	CE	68320	68350	68370	68321
325 012	**RM**	RM	*DB*	CE	68322	68351	68371	68323
325 013	**RM**	RM	*DB*	CE	68324	68352	68372	68325
325 014	**RM**	RM	*DB*	CE	68326	68353	68373	68327
325 015	**RM**	RM	*DB*	CE	68328	68354	68374	68329
325 016	**RM**	RM	*DB*	CE	68330	68355	68375	68331

Names (carried on one side of each DTPMV):

325 002   Royal Mail North Wales & North West
325 006   John Grierson
325 008   Peter Howarth CBE

CLASS 332   HEATHROW EXPRESS   SIEMENS

Dedicated Heathrow Express units. Five units were increased from 4-car to 5-car in 2002. Usually operate in coupled pairs.

Formations: Various.
Construction: Steel.
Traction Motors: Two Siemens monomotors asynchronous of 350 kW.
Wheel Arrangement: B-B + 2-2 + 2-2 (+ 2-2) + B-B.
Braking: Disc.
Dimensions: 23.74/23.35 x 2.75 m.
Bogies: CAF.
Couplers: Scharfenberg 10L.
Gangways: Within unit.
Control System: IGBT Inverter.
Doors: Sliding plug.
Maximum Speed: 100 mph.
Heating & ventilation: Air conditioning.
Seating: 1: 2+1 facing (* 1+1 facing/unidirectional), 2: 2+2 mainly unidirectional.
Multiple Working: Within class.

Note: A refurbishment programme is currently underway on these sets. Those refurbished so far are shown as *. First Class is being moved from one end to the other and the number of seats throughout the train are being changed. 72414 was the prototype refurbishment vehicle and at the time of writing was at Wolverton Works. It will be reinserted into 332 007 when that unit goes for refurbishment.

DMFO (*DMSO). CAF 1997–98. 26/– (* –/43 (8)). 48.8 t (* 49.9 t).
72400–413. TSO. CAF 1997–98. –/56 (* –/64 (11)). 35.8 t (* 38.4 t).
72414–418. TSO. CAF 2002. –/56 35.8 t.
PTSO. CAF 1997–98. –/44 1TD 1W (* –/39 (11) 1TD 2W). 45.6 t (* 47.6 t).
DMSO. CAF 1997–98. –/48. 48.8 t.

DMLFO. CAF 1997–98. 14/– 1W. 48.8 t.
DMFO (*). CAF 1997–98. 20/–.   . t.

332001–007. DMFO–TSO–PTSO–(TSO)–DMSO (* DMSO–TSO–PTSO–DMFO).
Advertising livery: Vehicles 78401, 78405, 78408, 78410, 78412 Vodafone (red).

332001		**HE**	HE	*HE*	OH	78400	72412	63400		78401
332002	*	**HE**	HE	*HE*	OH	78402	72409	63406		78403
332003	*	**HE**	HE	*HE*	OH	78404	72407	63402		78405
332004	*	**HE**	HE	*HE*	OH	78406	72405	63403		78407
332005		**HE**	HE	*HE*	OH	78408	72411	63404	72417	78409
332006		**HE**	HE	*HE*	OH	78410	72410	63405	72415	78411
332007		**HE**	HE	*HE*	OH	78412	72401	63401		78413
Spare		**HE**	HE		ZN				72414	

332008–014. DMSO–TSO–PTSO–(TSO)–DMLFO (* DMSO–TSO–PTSO–DMFO).
Advertising livery: Vehicles 78414, 78416, 78419, 78423, 78427 Vodafone (red).

332008		**HE**	HE	*HE*	OH	78414	72413	63407	72418	78415
332009		**HE**	HE	*HE*	OH	78416	72400	63408	72416	78417
332010	*	**HE**	HE	*HE*	OH	78418	72402	63409		78419
332011	*	**HE**	HE	*HE*	OH	78420	72403	63410		78421
332012		**HE**	HE	*HE*	OH	78422	72404	63411		78423
332013	*	**HE**	HE	*HE*	OH	78424	72408	63412		78425
332014		**HE**	HE	*HE*	OH	78426	72406	63413		78427

CLASS 333   SIEMENS

West Yorkshire area suburban units.

Formation: DMSO–PTSO–TSO–DMSO.
Construction: Steel.
Traction Motors: Two Siemens monomotors asynchronous of 350 kW.
Wheel Arrangement: B-B + 2-2 + 2-2 + B-B.
Braking: Disc.
Dimensions: 23.74 (outer ends)/23.35 (TSO) x 2.75 m.
Bogies: CAF.   **Couplers:** Dellner 10L.
Gangways: Within unit.   **Control System:** IGBT Inverter.
Doors: Sliding plug.   **Maximum Speed:** 100 mph.
Heating & ventilation: Air conditioning.
Seating Layout: 3+2 facing/unidirectional.
Multiple Working: Within class.

DMSO(A). (Odd Nos.) CAF 2001. –/90. 50.6 t.
PTSO. CAF 2001. –/73(6) 1TD 2W. 46.0 t.
TSO. CAF 2002–03. –/100. 38.5 t.
DMSO(B). (Even Nos.) CAF 2001. –/90. 50.0 t.

Notes: 333001–008 were made up to 4-car units from 3-car units in 2002.

333009–016 were made up to 4-car units from 3-car units in 2003.

333001	**YR**	A	*NO*	NL	78451	74461	74477	78452
333002	**YR**	A	*NO*	NL	78453	74462	74478	78454

333003	**YR**	A	*NO*	NL	78455	74463	74479	78456
333004	**YR**	A	*NO*	NL	78457	74464	74480	78458
333005	**YR**	A	*NO*	NL	78459	74465	74481	78460
333006	**YR**	A	*NO*	NL	78461	74466	74482	78462
333007	**YR**	A	*NO*	NL	78463	74467	74483	78464
333008	**YR**	A	*NO*	NL	78465	74468	74484	78466
333009	**YR**	A	*NO*	NL	78467	74469	74485	78468
333010	**YR**	A	*NO*	NL	78469	74470	74486	78470
333011	**YR**	A	*NO*	NL	78471	74471	74487	78472
333012	**YR**	A	*NO*	NL	78473	74472	74488	78474
333013	**YR**	A	*NO*	NL	78475	74473	74489	78476
333014	**YR**	A	*NO*	NL	78477	74474	74490	78478
333015	**YR**	A	*NO*	NL	78479	74475	74491	78480
333016	**YR**	A	*NO*	NL	78481	74476	74492	78482

Name (carried on end cars):

333007   Alderman J Arthur Godwin First Lord Mayor of Bradford 1907

CLASS 334   JUNIPER   ALSTOM BIRMINGHAM

Outer suburban units.

Formation: DMSO–PTSO–DMSO.
Construction: Steel.
Traction Motors: Two Alstom ONIX 800 asynchronous of 270 kW.
Wheel Arrangement: 2-Bo + 2-2 + Bo-2.
Braking: Disc.   **Dimensions:** 21.01/19.94 x 2.80 m.
Bogies: Alstom LTB3/TBP3.   **Couplers:** Tightlock.
Gangways: Within unit.   **Control System:** IGBT Inverter.
Doors: Sliding plug.   **Maximum Speed:** 90 mph.
Heating & ventilation: Pressure heating and ventilation.
Seating Layout: 2+2 facing/unidirectional (3+2 in PTSO).
Multiple Working: Within class.

DMSO(A). Alstom Birmingham 1999–2001. –/64. 42.6 t.
PTSO. Alstom Birmingham 1999–2001. –/55 1TD 1W. 39.4 t.
DMSO(B). Alstom Birmingham 1999–2001. –/64. 42.6 t.

334001	**SP**	E	*SR*	GW	64101	74301	65101	Donald Dewar
334002	**SP**	E	*SR*	GW	64102	74302	65102	
334003	**SP**	E	*SR*	GW	64103	74303	65103	
334004	**SP**	E	*SR*	GW	64104	74304	65104	
334005	**SP**	E	*SR*	GW	64105	74305	65105	
334006	**SR**	E	*SR*	GW	64106	74306	65106	
334007	**SP**	E	*SR*	GW	64107	74307	65107	
334008	**SP**	E	*SR*	GW	64108	74308	65108	
334009	**SP**	E	*SR*	GW	64109	74309	65109	
334010	**SP**	E	*SR*	GW	64110	74310	65110	
334011	**SP**	E	*SR*	GW	64111	74311	65111	
334012	**SR**	E	*SR*	GW	64112	74312	65112	
334013	**SP**	E	*SR*	GW	64113	74313	65113	
334014	**SP**	E	*SR*	GW	64114	74314	65114	

334015	**SP**	E	*SR*	GW	64115	74315	65115	
334016	**SP**	E	*SR*	GW	64116	74316	65116	
334017	**SP**	E	*SR*	GW	64117	74317	65117	
334018	**SP**	E	*SR*	GW	64118	74318	65118	
334019	**SP**	E	*SR*	GW	64119	74319	65119	
334020	**SR**	E	*SR*	GW	64120	74320	65120	
334021	**SP**	E	*SR*	GW	64121	74321	65121	Larkhall
334022	**SP**	E	*SR*	GW	64122	74322	65122	
334023	**SR**	E	*SR*	GW	64123	74323	65123	
334024	**SP**	E	*SR*	GW	64124	74324	65124	
334025	**SP**	E	*SR*	GW	64125	74325	65125	
334026	**SP**	E	*SR*	GW	64126	74326	65126	
334027	**SR**	E	*SR*	GW	64127	74327	65127	
334028	**SR**	E	*SR*	GW	64128	74328	65128	
334029	**SR**	E	*SR*	GW	64129	74329	65129	
334030	**SP**	E	*SR*	GW	64130	74330	65130	
334031	**SR**	E	*SR*	GW	64131	74331	65131	
334032	**SP**	E	*SR*	GW	64132	74332	65132	
334033	**SP**	E	*SR*	GW	64133	74333	65133	
334034	**SP**	E	*SR*	GW	64134	74334	65134	
334035	**SP**	E	*SR*	GW	64135	74335	65135	
334036	**SP**	E	*SR*	GW	64136	74336	65136	
334037	**SP**	E	*SR*	GW	64137	74337	65137	
334038	**SR**	E	*SR*	GW	64138	74338	65138	
334039	**SP**	E	*SR*	GW	64139	74339	65139	
334040	**SP**	E	*SR*	GW	64140	74340	65140	

CLASS 350   DESIRO UK   SIEMENS

Outer suburban and long distance units.

Formation: DMCO–TCO–PTSO–DMCO.
Systems: 25 kV AC overhead (350/1s built with 750 V DC).
Construction: Welded aluminium.
Traction Motors: 4 Siemens 1TB2016-0GB02 asynchronous of 250 kW.
Wheel Arrangement: Bo-Bo + 2-2 + 2-2 + Bo-Bo.
Braking: Disc & regenerative.   **Dimensions:** 20.34 x 2.79 m.
Bogies: SGP SF5000.   **Couplers:** Dellner 12.
Gangways: Throughout.   **Control System:** IGBT Inverter.
Doors: Sliding plug.
Maximum Speed: 100 mph (max. speed of 350/1s being increased to 110 mph).
Heating & ventilation: Air conditioning.
Seating Layout: 1: 2+2 facing, 2: 2+2 facing/unidirectional (3+2 in 350/2s).
Multiple Working: Within class.

Class 350/1. Original build units owned by Angel Trains. Formerly part of an aborted South West Trains 5-car Class 450/2 order. 2+2 seating.

DMSO(A). Siemens Krefeld 2004–05. –/60. 48.7 t.
TCO. Siemens Krefeld/Prague 2004–05. 24/32 1T. 36.2 t.
PTSO. Siemens Krefeld/Prague 2004–05. –/50(9) 1TD 2W. 45.2 t.
DMSO(B). Siemens Krefeld 2004–05. –/60. 49.2 t.

350 101	**LM**	A	*LM*	NN	63761	66811	66861	63711
350 102	**LM**	A	*LM*	NN	63762	66812	66862	63712
350 103	**LM**	A	*LM*	NN	63765	66813	66863	63713
350 104	**LM**	A	*LM*	NN	63764	66814	66864	63714
350 105	**LM**	A	*LM*	NN	63763	66815	66868	63715
350 106	**LM**	A	*LM*	NN	63766	66816	66866	63716
350 107	**LM**	A	*LM*	NN	63767	66817	66867	63717
350 108	**LM**	A	*LM*	NN	63768	66818	66865	63718
350 109	**LM**	A	*LM*	NN	63769	66819	66869	63719
350 110	**LM**	A	*LM*	NN	63770	66820	66870	63720
350 111	**LM**	A	*LM*	NN	63771	66821	66871	63721
350 112	**LM**	A	*LM*	NN	63772	66822	66872	63722
350 113	**LM**	A	*LM*	NN	63773	66823	66873	63723
350 114	**LM**	A	*LM*	NN	63774	66824	66874	63724
350 115	**LM**	A	*LM*	NN	63775	66825	66875	63725
350 116	**LM**	A	*LM*	NN	63776	66826	66876	63726
350 117	**LM**	A	*LM*	NN	63777	66827	66877	63727
350 118	**LM**	A	*LM*	NN	63778	66828	66878	63728
350 119	**LM**	A	*LM*	NN	63779	66829	66879	63729
350 120	**LM**	A	*LM*	NN	63780	66830	66880	63730
350 121	**LM**	A	*LM*	NN	63781	66831	66881	63731
350 122	**LM**	A	*LM*	NN	63782	66832	66882	63732
350 123	**LM**	A	*LM*	NN	63783	66833	66883	63733
350 124	**LM**	A	*LM*	NN	63784	66834	66884	63734
350 125	**LM**	A	*LM*	NN	63785	66835	66885	63735
350 126	**LM**	A	*LM*	NN	63786	66836	66886	63736
350 127	**LM**	A	*LM*	NN	63787	66837	66887	63737
350 128	**LM**	A	*LM*	NN	63788	66838	66888	63738
350 129	**LM**	A	*LM*	NN	63789	66839	66889	63739
350 130	**LM**	A	*LM*	NN	63790	66840	66890	63740

Class 350/2. Owned by Porterbrook Leasing. 3+2 seating.

DMSO(A). Siemens Krefeld 2008–09. –/70. 43.7 t.
TCO. Siemens Prague 2008–09. 24/42 1T. 35.3 t.
PTSO. Siemens Prague 2008–09. –/61(9) 1TD 2W. 42.9 t.
DMSO(B). Siemens Krefeld 2008–09. –/70. 44.2 t.

350 231	**LM**	P	*LM*	NN	61431	65231	67531	61531
350 232	**LM**	P	*LM*	NN	61432	65232	67532	61532
350 233	**LM**	P	*LM*	NN	61433	65233	67533	61533
350 234	**LM**	P	*LM*	NN	61434	65234	67534	61534
350 235	**LM**	P	*LM*	NN	61435	65235	67535	61535
350 236	**LM**	P	*LM*	NN	61436	65236	67536	61536
350 237	**LM**	P	*LM*	NN	61437	65237	67537	61537
350 238	**LM**	P	*LM*	NN	61438	65238	67538	61538
350 239	**LM**	P	*LM*	NN	61439	65239	67539	61539
350 240	**LM**	P	*LM*	NN	61440	65240	67540	61540
350 241	**LM**	P	*LM*	NN	61441	65241	67541	61541
350 242	**LM**	P	*LM*	NN	61442	65242	67542	61542
350 243	**LM**	P	*LM*	NN	61443	65243	67543	61543
350 244	**LM**	P	*LM*	NN	61444	65244	67544	61544

350245	**LM**	P	*LM*	NN	61445	65245	67545	61545
350246	**LM**	P	*LM*	NN	61446	65246	67546	61546
350247	**LM**	P	*LM*	NN	61447	65247	67547	61547
350248	**LM**	P	*LM*	NN	61448	65248	67548	61548
350249	**LM**	P	*LM*	NN	61449	65249	67549	61549
350250	**LM**	P	*LM*	NN	61450	65250	67550	61550
350251	**LM**	P	*LM*	NN	61451	65251	67551	61551
350252	**LM**	P	*LM*	NN	61452	65252	67552	61552
350253	**LM**	P	*LM*	NN	61453	65253	67553	61553
350254	**LM**	P	*LM*	NN	61454	65254	67554	61554
350255	**LM**	P	*LM*	NN	61455	65255	67555	61555
350256	**LM**	P	*LM*	NN	61456	65256	67556	61556
350257	**LM**	P	*LM*	NN	61457	65257	67557	61557
350258	**LM**	P	*LM*	NN	61458	65258	67558	61558
350259	**LM**	P	*LM*	NN	61459	65259	67559	61559
350260	**LM**	P	*LM*	NN	61460	65260	67560	61560
350261	**LM**	P	*LM*	NN	61461	65261	67561	61561
350262	**LM**	P	*LM*	NN	61462	65262	67562	61562
350263	**LM**	P	*LM*	NN	61463	65263	67563	61563
350264	**LM**	P	*LM*	NN	61464	65264	67564	61564
350265	**LM**	P	*LM*	NN	61465	65265	67565	61565
350266	**LM**	P	*LM*	NN	61466	65266	67566	61566
350267	**LM**	P	*LM*	NN	61467	65267	67567	61567

Name (carried on PTSO): 350 232   Chad Varah

Class 350/3. Owned by Angel Trains. On order for London Midland and due for delivery 2014. Full details awaited; numbering series is provisional.

DMSO(A). Siemens Krefeld 2013–14.   .   t.
TCO. Siemens Krefeld 2013–14.   .   t.
PTSO. Siemens Krefeld 2013–14.   .   t.
DMSO(B). Siemens Krefeld 2013–14.   .   t.

350368	A	60141	60511	60651	60151
350369	A	60142	60512	60652	60152
350370	A	60143	60513	60653	60153
350371	A	60144	60514	60654	60154
350372	A	60145	60515	60655	60155
350373	A	60146	60516	60656	60156
350374	A	60147	60517	60657	60157
350375	A	60148	60518	60658	60158
350376	A	60149	60519	60659	60159
350377	A	60150	60520	60660	60160

Class 350/4. Owned by Angel Trains. Under construction for TransPennine Express Manchester Airport–Edinburgh/Glasgow services and due for delivery 2013–14. Full details awaited; numbering series is provisional.

DMSO(A). Siemens Krefeld 2012–13.   .   t.
TCO. Siemens Krefeld 2012–13.   .   t.
PTSO. Siemens Krefeld 2012–13.   .   t.
DMSO(B). Siemens Krefeld 2012–13.   .   t.

350401	A	60651	60901	60941	60671
350402	A	60652	60902	60942	60672
350403	A	60653	60903	60943	60673
350404	A	60654	60904	60944	60674
350405	A	60655	60905	60945	60675
350406	A	60656	60906	60946	60676
350407	A	60657	60907	60947	60677
350408	A	60658	60908	60948	60678
350409	A	60659	60909	60949	60679
350410	A	60660	60910	60950	60680

CLASS 357 ELECTROSTAR
ADTRANZ/BOMBARDIER DERBY

Provision for 750 V DC supply if required.

Formation: DMSO–MSO–PTSO–DMSO.
Construction: Welded aluminium alloy underframe, sides and roof with steel ends. All sections bolted together.
Traction Motors: Two Adtranz asynchronous of 250 kW.
Wheel Arrangement: 2-Bo + 2-Bo + 2-2 + Bo-2.
Braking: Disc & regenerative.   **Dimensions:** 20.40/19.99 x 2.80 m.
Bogies: Adtranz P3-25/T3-25.   **Couplers:** Tightlock.
Gangways: Within unit.   **Control System:** IGBT Inverter.
Doors: Sliding plug.   **Maximum Speed:** 100 mph.
Heating & ventilation: Air conditioning.
Seating Layout: 3+2 facing/unidirectional.
Multiple Working: Within class.

Class 357/0. Owned by Porterbrook Leasing.

DMSO(A). Adtranz Derby 1999–2001. –/71. 40.7 t.
MSO. Adtranz Derby 1999–2001. –/78. 36.7 t.
PTSO. Adtranz Derby 1999–2001. –/58(4) 1TD 2W. 39.5 t.
DMSO(B). Adtranz Derby 1999–2001. –/71. 40.7 t.

357001	**NC**	P	*C2*	EM	67651	74151	74051	67751
357002	**NC**	P	*C2*	EM	67652	74152	74052	67752
357003	**NC**	P	*C2*	EM	67653	74153	74053	67753
357004	**NC**	P	*C2*	EM	67654	74154	74054	67754
357005	**NC**	P	*C2*	EM	67655	74155	74055	67755
357006	**NC**	P	*C2*	EM	67656	74156	74056	67756
357007	**NC**	P	*C2*	EM	67657	74157	74057	67757
357008	**NC**	P	*C2*	EM	67658	74158	74058	67758
357009	**NC**	P	*C2*	EM	67659	74159	74059	67759
357010	**NC**	P	*C2*	EM	67660	74160	74060	67760
357011	**NC**	P	*C2*	EM	67661	74161	74061	67761
357012	**NC**	P	*C2*	EM	67662	74162	74062	67762
357013	**NC**	P	*C2*	EM	67663	74163	74063	67763
357014	**NC**	P	*C2*	EM	67664	74164	74064	67764
357015	**NC**	P	*C2*	EM	67665	74165	74065	67765
357016	**NC**	P	*C2*	EM	67666	74166	74066	67766

357017	**NC**	P	*C2*	EM	67667	74167	74067	67767
357018	**NC**	P	*C2*	EM	67668	74168	74068	67768
357019	**NC**	P	*C2*	EM	67669	74169	74069	67769
357020	**NC**	P	*C2*	EM	67670	74170	74070	67770
357021	**NC**	P	*C2*	EM	67671	74171	74071	67771
357022	**NC**	P	*C2*	EM	67672	74172	74072	67772
357023	**NC**	P	*C2*	EM	67673	74173	74073	67773
357024	**NC**	P	*C2*	EM	67674	74174	74074	67774
357025	**NC**	P	*C2*	EM	67675	74175	74075	67775
357026	**NC**	P	*C2*	EM	67676	74176	74076	67776
357027	**NC**	P	*C2*	EM	67677	74177	74077	67777
357028	**NC**	P	*C2*	EM	67678	74178	74078	67778
357029	**NC**	P	*C2*	EM	67679	74179	74079	67779
357030	**NC**	P	*C2*	EM	67680	74180	74080	67780
357031	**NC**	P	*C2*	EM	67681	74181	74081	67781
357032	**NC**	P	*C2*	EM	67682	74182	74082	67782
357033	**NC**	P	*C2*	EM	67683	74183	74083	67783
357034	**NC**	P	*C2*	EM	67684	74184	74084	67784
357035	**NC**	P	*C2*	EM	67685	74185	74085	67785
357036	**NC**	P	*C2*	EM	67686	74186	74086	67786
357037	**NC**	P	*C2*	EM	67687	74187	74087	67787
357038	**NC**	P	*C2*	EM	67688	74188	74088	67788
357039	**NC**	P	*C2*	EM	67689	74189	74089	67789
357040	**NC**	P	*C2*	EM	67690	74190	74090	67790
357041	**NC**	P	*C2*	EM	67691	74191	74091	67791
357042	**NC**	P	*C2*	EM	67692	74192	74092	67792
357043	**NC**	P	*C2*	EM	67693	74193	74093	67793
357044	**NC**	P	*C2*	EM	67694	74194	74094	67794
357045	**NC**	P	*C2*	EM	67695	74195	74095	67795
357046	**NC**	P	*C2*	EM	67696	74196	74096	67796

Names (carried on DMSO(A) and DMSO(B) (one plate on each)):

357001 BARRY FLAXMAN
357002 ARTHUR LEWIS STRIDE 1841–1922
357003 SOUTHEND city.on.sea
357004 TONY AMOS
357016 Diamond Jubilee 1952–2012
357011 JOHN LOWING
357028 London, Tilbury & Southend Railway 1854–2004
357029 THOMAS WHITELEGG 1840–1922
357030 ROBERT HARBEN WHITELEGG 1871–1957

Class 357/2. Owned by Angel Trains.

DMSO(A). Bombardier Derby 2001–02. –/71. 40.7 t.
MSO. Bombardier Derby 2001–02. –/78. 36.7 t.
PTSO. Bombardier Derby 2001–02. –/58(4) 1TD 2W. 39.5 t.
DMSO(B). Bombardier Derby 2001–02. –/71. 40.7 t.

357201	**NC**	A	*C2*	EM	68601	74701	74601	68701
357202	**NC**	A	*C2*	EM	68602	74702	74602	68702
357203	**NC**	A	*C2*	EM	68603	74703	74603	68703

357 204	**NC**	A	*C2*	EM	68604	74704	74604	68704
357 205	**NC**	A	*C2*	EM	68605	74705	74605	68705
357 206	**NC**	A	*C2*	EM	68606	74706	74606	68706
357 207	**NC**	A	*C2*	EM	68607	74707	74607	68707
357 208	**NC**	A	*C2*	EM	68608	74708	74608	68708
357 209	**NC**	A	*C2*	EM	68609	74709	74609	68709
357 210	**NC**	A	*C2*	EM	68610	74710	74610	68710
357 211	**NC**	A	*C2*	EM	68611	74711	74611	68711
357 212	**NC**	A	*C2*	EM	68612	74712	74612	68712
357 213	**NC**	A	*C2*	EM	68613	74713	74613	68713
357 214	**NC**	A	*C2*	EM	68614	74714	74614	68714
357 215	**NC**	A	*C2*	EM	68615	74715	74615	68715
357 216	**NC**	A	*C2*	EM	68616	74716	74616	68716
357 217	**NC**	A	*C2*	EM	68617	74717	74617	68717
357 218	**NC**	A	*C2*	EM	68618	74718	74618	68718
357 219	**NC**	A	*C2*	EM	68619	74719	74619	68719
357 220	**NC**	A	*C2*	EM	68620	74720	74620	68720
357 221	**NC**	A	*C2*	EM	68621	74721	74621	68721
357 222	**NC**	A	*C2*	EM	68622	74722	74622	68722
357 223	**NC**	A	*C2*	EM	68623	74723	74623	68723
357 224	**NC**	A	*C2*	EM	68624	74724	74624	68724
357 225	**NC**	A	*C2*	EM	68625	74725	74625	68725
357 226	**NC**	A	*C2*	EM	68626	74726	74626	68726
357 227	**NC**	A	*C2*	EM	68627	74727	74627	68727
357 228	**NC**	A	*C2*	EM	68628	74728	74628	68728

Names (carried on DMSO(A) and DMSO(B) (one plate on each)):

357 201	KEN BIRD	357 207	JOHN PAGE
357 202	KENNY MITCHELL	357 208	DAVE DAVIS
357 203	HENRY PUMFRETT	357 209	JAMES SNELLING
357 204	DEREK FOWERS	357 213	UPMINSTER I.E.C.C.
357 205	JOHN D'SILVA	357 217	ALLAN BURNELL
357 206	MARTIN AUNGIER		

CLASS 360/0   DESIRO UK   SIEMENS

Outer suburban/express units.

Formation: DMCO–PTSO–TSO–DMCO.
Construction: Welded aluminium.
Traction Motors: 4 Siemens 1TB2016-0GB02 asynchronous of 250 kW.
Wheel Arrangement: Bo-Bo + 2-2 + 2-2 + Bo-Bo.
Braking: Disc & regenerative.   **Dimensions:** 20.34 x 2.80 m.
Bogies: SGP SF5000.   **Couplers:** Dellner 12.
Gangways: Within unit.   **Control System:** IGBT Inverter.
Doors: Sliding plug.   **Maximum Speed:** 100 mph.
Heating & ventilation: Air conditioning.
Seating Layout: 1: 2+2 facing, 2: 3+2 facing/unidirectional.
Multiple Working: Within class.

DMCO(A). Siemens Krefeld 2002–03. 8/59. 45.0 t.
PTSO. Siemens Vienna 2002–03. –/60(9) 1TD 2W. 43.0 t.
TSO. Siemens Vienna 2002–03. –/78. 35.0 t.
DMCO(B). Siemens Krefeld 2002–03. 8/59. 45.0 t.

360 101	**FB**	A	*EA*	IL	65551	72551	74551	68551
360 102	**FB**	A	*EA*	IL	65552	72552	74552	68552
360 103	**FB**	A	*EA*	IL	65553	72553	74553	68553
360 104	**FB**	A	*EA*	IL	65554	72554	74554	68554
360 105	**FB**	A	*EA*	IL	65555	72555	74555	68555
360 106	**FB**	A	*EA*	IL	65556	72556	74556	68556
360 107	**FB**	A	*EA*	IL	65557	72557	74557	68557
360 108	**FB**	A	*EA*	IL	65558	72558	74558	68558
360 109	**FB**	A	*EA*	IL	65559	72559	74559	68559
360 110	**FB**	A	*EA*	IL	65560	72560	74560	68560
360 111	**FB**	A	*EA*	IL	65561	72561	74561	68561
360 112	**FB**	A	*EA*	IL	65562	72562	74562	68562
360 113	**FB**	A	*EA*	IL	65563	72563	74563	68563
360 114	**FB**	A	*EA*	IL	65564	72564	74564	68564
360 115	**FB**	A	*EA*	IL	65565	72565	74565	68565
360 116	**FB**	A	*EA*	IL	65566	72566	74566	68566
360 117	**FB**	A	*EA*	IL	65567	72567	74567	68567
360 118	**FB**	A	*EA*	IL	65568	72568	74568	68568
360 119	**FB**	A	*EA*	IL	65569	72569	74569	68569
360 120	**FB**	A	*EA*	IL	65570	72570	74570	68570
360 121	**FB**	A	*EA*	IL	65571	72571	74571	68571

CLASS 360/2   DESIRO UK   SIEMENS

4-car Class 350 testbed units rebuilt for use by Heathrow Express on Paddington–Heathrow Airport stopping services ("Heathrow Connect").

Original 4-car sets 360 201–204 were made up to 5-cars during 2007 using additional TSOs. A fifth unit (360 205) was delivered in late 2005 as a 5-car set. This set is now dedicated to Terminals 1&3–Terminal 4 shuttle services.

Formation: DMSO–PTSO–TSO–TSO–DMSO.
Construction: Welded aluminium.
Traction Motors: 4 Siemens 1TB2016-0GB02 asynchronous of 250 kW.
Wheel Arrangement: Bo-Bo + 2-2 + 2-2 + 2-2 + Bo-Bo.

Braking: Disc & regenerative.	**Dimensions:** 20.34 x 2.80 m.
Bogies: SGP SF5000.	**Couplers:** Dellner 12.
Gangways: Within unit.	**Control System:** IGBT Inverter.
Doors: Sliding plug.	**Maximum Speed:** 100 mph.

Heating & ventilation: Air conditioning.
Seating Layout: 3+2 (* 2+2) facing/unidirectional.
Multiple Working: Within class.

DMSO(A). Siemens Krefeld 2002–06. –/63 (* –/54). 44.8 t.
PTSO. Siemens Krefeld 2002–06. –/57(9) 1TD 2W (* –/48(9) 2W). 44.2 t.
TSO. Siemens Krefeld 2005–06. –/74 (* –/62). 35.3 t.
TSO. Siemens Krefeld 2002–06. –/74 (* –/62). 34.1 t.
DMSO(B). Siemens Krefeld 2002–06. –/63 (* –/54). 44.4 t.

360 201		**HC**	HE	*HC*	OH	78431	63421	72431	72421	78441
360 202		**HC**	HE	*HC*	OH	78432	63422	72432	72422	78442
360 203		**HC**	HE	*HC*	OH	78433	63423	72433	72423	78443
360 204		**HC**	HE	*HC*	OH	78434	63424	72434	72424	78444
360 205	*	**HE**	HE	*HE*	OH	78435	63425	72435	72425	78445

CLASS 365　　　NETWORKER EXPRESS　　　ABB YORK

Outer suburban units.

Formations: DMCO–TSO–PTSO–DMCO.
Systems: 25 kV AC overhead but with 750 V DC third rail capability (units marked * were formerly used on DC lines in the South-East).
Construction: Welded aluminium alloy.
Traction Motors: Four GEC-Alsthom G354CX asynchronous of 157 kW.
Wheel Arrangement: Bo-Bo + 2-2 + 2-2 + Bo-Bo.
Braking: Disc & rheostatic.
Dimensions: 20.89/20.06 x 2.81 m.
Bogies: ABB P3-16/T3-16.　　　　　　　　**Couplers:** Tightlock.
Gangways: Within unit.　　　　　　　　　**Control System:** GTO Inverter.
Doors: Sliding plug.　　　　　　　　　　**Maximum Speed:** 100 mph.
Seating Layout: 1: 2+2 facing, 2: 2+2 facing.
Multiple Working: Within class only.

DMCO(A). Lot No. 31133 1994–95. 12/56. 41.7 t.
TSO. Lot No. 31134 1994–95. –/65 1TD (* –/64 1TD) 32.9 t.
PTSO. Lot No. 31135 1994–95. –/68 1T. 34.6 t.
DMCO(B). Lot No. 31136 1994–95. 12/56. 41.7 t.

Advertising liveries:

365510 Cambridge & Ely; Cathedral cities (blue & white with various images).
365519 Peterborough; environment capital (blue & white with various images).
365531 Nelson's County; Norfolk (blue & white with various images).
365540 Garden cities of Hertfordshire (blue & white with various images).

365501	*	**FU**	E	*FC*	HE	65894	72241	72240	65935
365502	*	**FU**	E	*FC*	HE	65895	72243	72242	65936
365503	*	**FU**	E	*FC*	HE	65896	72245	72244	65937
365504	*	**FU**	E	*FC*	HE	65897	72247	72246	65938
365505	*	**FU**	E	*FC*	HE	65898	72249	72248	65939
365506	*	**FU**	E	*FC*	HE	65899	72251	72250	65940
365507	*	**FU**	E	*FC*	HE	65900	72253	72252	65941
365508	*	**FU**	E	*FC*	HE	65901	72255	72254	65942
365509	*	**FU**	E	*FC*	HE	65902	72257	72256	65943
365510	*	**AL**	E	*FC*	HE	65903	72259	72258	65944
365511	*	**FU**	E	*FC*	HE	65904	72261	72260	65945
365512	*	**FU**	E	*FC*	HE	65905	72263	72262	65946
365513	*	**FU**	E	*FC*	HE	65906	72265	72264	65947
365514	*	**FU**	E	*FC*	HE	65907	72267	72266	65948
365515	*	**FU**	E	*FC*	HE	65908	72269	72268	65949
365516	*	**FU**	E	*FC*	HE	65909	72271	72270	65950
365517		**FU**	E	*FC*	HE	65910	72273	72272	65951

365518	**FU**	E	*FC*	HE	65911	72275	72274	65952
365519	**AL**	E	*FC*	HE	65912	72277	72276	65953
365520	**FU**	E	*FC*	HE	65913	72279	72278	65954
365521	**FU**	E	*FC*	HE	65914	72281	72280	65955
365522	**FU**	E	*FC*	HE	65915	72283	72282	65956
365523	**FU**	E	*FC*	HE	65916	72285	72284	65957
365524	**FU**	E	*FC*	HE	65917	72287	72286	65958
365525	**FU**	E	*FC*	HE	65918	72289	72288	65959
365527	**FU**	E	*FC*	HE	65920	72293	72292	65961
365528	**FU**	E	*FC*	HE	65921	72295	72294	65962
365529	**FU**	E	*FC*	HE	65922	72297	72296	65963
365530	**FU**	E	*FC*	HE	65923	72299	72298	65964
365531	**AL**	E	*FC*	HE	65924	72301	72300	65965
365532	**FU**	E	*FC*	HE	65925	72303	72302	65966
365533	**FU**	E	*FC*	HE	65926	72305	72304	65967
365534	**FU**	E	*FC*	HE	65927	72307	72306	65968
365535	**FU**	E	*FC*	HE	65928	72309	72308	65969
365536	**FU**	E	*FC*	HE	65929	72311	72310	65970
365537	**FU**	E	*FC*	HE	65930	72313	72312	65971
365538	**FU**	E	*FC*	HE	65931	72315	72314	65972
365539	**FU**	E	*FC*	HE	65932	72317	72316	65973
365540	**AL**	E	*FC*	HE	65933	72319	72318	65974
365541	**FU**	E	*FC*	HE	65934	72321	72320	65975

Names (carried on each DMCO):

365506 The Royston Express
365513 Hornsey Depot
365514 Captain George Vancouver
365518 The Fenman
365527 Robert Stripe Passengers' Champion
365530 The Intalink Partnership promoting integrated transport in
        Hertfordshire since 1999
365536 Rufus Barnes Chief Executive of London TravelWatch for 25 years
365537 Daniel Edwards (1974–2010) Cambridge Driver

CLASS 375 ELECTROSTAR
ADTRANZ/BOMBARDIER DERBY

Express and outer suburban units.

Formations: Various.
Systems: 25 kV AC overhead/750 V DC third rail (some third rail only with provision for retro-fitting of AC equipment).
Construction: Welded aluminium alloy underframe, sides and roof with steel ends. All sections bolted together.
Traction Motors: Two Adtranz asynchronous of 250 kW.
Wheel Arrangement: 2-Bo (+ 2-Bo) + 2-2 + Bo-2.

Braking: Disc & regenerative.	**Dimensions:** 20.40/19.99 x 2.80 m.
Bogies: Adtranz P3-25/T3-25.	**Couplers:** Dellner 12.
Gangways: Throughout.	**Control System:** IGBT Inverter.

Doors: Sliding plug.   **Maximum Speed:** 100 mph.
Heating & ventilation: Air conditioning.
Seating Layout: 1: 2+2 facing/unidirectional (seats behind drivers cab in each DMCO). 2: 2+2 facing/unidirectional (except 375/9 – 3+2 facing/unidirectional).
Multiple Working: Within class and with Classes 376, 377, 378 and 379.

Class 375/3. Express units. 750 V DC only. DMCO–TSO–DMCO.

DMCO(A). Bombardier Derby 2001–02. 12/48. 43.8 t.
TSO. Bombardier Derby 2001–02. –/56 1TD 2W. 35.5 t.
DMCO(B). Bombardier Derby 2001–02. 12/48. 43.8 t.

375301	**CN**	E	*SE*	RM	67921	74351	67931
375302	**CN**	E	*SE*	RM	67922	74352	67932
375303	**CN**	E	*SE*	RM	67923	74353	67933
375304	**CN**	E	*SE*	RM	67924	74354	67934
375305	**CN**	E	*SE*	RM	67925	74355	67935
375306	**CN**	E	*SE*	RM	67926	74356	67936
375307	**CN**	E	*SE*	RM	67927	74357	67937
375308	**SE**	E	*SE*	RM	67928	74358	67938
375309	**CN**	E	*SE*	RM	67929	74359	67939
375310	**CN**	E	*SE*	RM	67930	74360	67940

Name (carried on TSO):

375304 Medway Valley Line 1856–2006

Class 375/6. Express units. 25 kV AC/750 V DC. DMCO–MSO–PTSO–DMCO.

DMCO(A). Adtranz Derby 1999–2001. 12/48. 46.2 t.
MSO. Adtranz Derby 1999–2001. –/66 1T. 40.5 t.
PTSO. Adtranz Derby 1999–2001. –/56 1TD 2W. 40.7 t.
DMCO(B). Adtranz Derby 1999–2001. 12/48. 46.2 t.

375601	**CN**	E	*SE*	RM	67801	74251	74201	67851
375602	**CN**	E	*SE*	RM	67802	74252	74202	67852
375603	**CN**	E	*SE*	RM	67803	74253	74203	67853
375604	**CN**	E	*SE*	RM	67804	74254	74204	67854
375605	**CN**	E	*SE*	RM	67805	74255	74205	67855
375606	**CN**	E	*SE*	RM	67806	74256	74206	67856
375607	**CN**	E	*SE*	RM	67807	74257	74207	67857
375608	**CN**	E	*SE*	RM	67808	74258	74208	67858
375609	**CN**	E	*SE*	RM	67809	74259	74209	67859
375610	**CN**	E	*SE*	RM	67810	74260	74210	67860
375611	**CN**	E	*SE*	RM	67811	74261	74211	67861
375612	**CN**	E	*SE*	RM	67812	74262	74212	67862
375613	**CN**	E	*SE*	RM	67813	74263	74213	67863
375614	**CN**	E	*SE*	RM	67814	74264	74214	67864
375615	**CN**	E	*SE*	RM	67815	74265	74215	67865
375616	**CN**	E	*SE*	RM	67816	74266	74216	67866
375617	**CN**	E	*SE*	RM	67817	74267	74217	67867
375618	**CN**	E	*SE*	RM	67818	74268	74218	67868
375619	**CN**	E	*SE*	RM	67819	74269	74219	67869
375620	**CN**	E	*SE*	RM	67820	74270	74220	67870
375621	**CN**	E	*SE*	RM	67821	74271	74221	67871

375622	**CN**	E	*SE*	RM	67822	74272	74222	67872
375623	**CN**	E	*SE*	RM	67823	74273	74223	67873
375624	**SE**	E	*SE*	RM	67824	74274	74224	67874
375625	**CN**	E	*SE*	RM	67825	74275	74225	67875
375626	**CN**	E	*SE*	RM	67826	74276	74226	67876
375627	**CN**	E	*SE*	RM	67827	74277	74227	67877
375628	**CN**	E	*SE*	RM	67828	74278	74228	67878
375629	**CN**	E	*SE*	RM	67829	74279	74229	67879
375630	**CN**	E	*SE*	RM	67830	74280	74230	67880

Names (carried on one side of each MSO or PTSO):

375608 Bromley Travelwise	375619 Driver John Neve
375610 Royal Tunbridge Wells	375623 Hospice in the Weald
375611 Dr. William Harvey	

Class 375/7. Express units. 750 V DC only. DMCO–MSO–TSO–DMCO.

DMCO(A). Bombardier Derby 2001–02. 12/48. 43.8 t.
MSO. Bombardier Derby 2001–02. –/66 1T. 36.4 t.
TSO. Bombardier Derby 2001–02. –/56 1TD 2W. 34.1 t.
DMCO(B). Bombardier Derby 2001–02. 12/48. 43.8 t.

375701	**CN**	E	*SE*	RM	67831	74281	74231	67881
375702	**CN**	E	*SE*	RM	67832	74282	74232	67882
375703	**CN**	E	*SE*	RM	67833	74283	74233	67883
375704	**CN**	E	*SE*	RM	67834	74284	74234	67884
375705	**SE**	E	*SE*	RM	67835	74285	74235	67885
375706	**CN**	E	*SE*	RM	67836	74286	74236	67886
375707	**CN**	E	*SE*	RM	67837	74287	74237	67887
375708	**CN**	E	*SE*	RM	67838	74288	74238	67888
375709	**CN**	E	*SE*	RM	67839	74289	74239	67889
375710	**CN**	E	*SE*	RM	67840	74290	74240	67890
375711	**CN**	E	*SE*	RM	67841	74291	74241	67891
375712	**CN**	E	*SE*	RM	67842	74292	74242	67892
375713	**CN**	E	*SE*	RM	67843	74293	74243	67893
375714	**CN**	E	*SE*	RM	67844	74294	74244	67894
375715	**CN**	E	*SE*	RM	67845	74295	74245	67895

Name (carried on one side of each MSO or TSO):

375701 Kent Air Ambulance Explorer

Class 375/8. Express units. 750 V DC only. DMCO–MSO–TSO–DMCO.

DMCO(A). Bombardier Derby 2004. 12/48. 43.3 t.
MSO. Bombardier Derby 2004. –/66 1T. 39.8 t.
TSO. Bombardier Derby 2004. –/52 1TD 2W. 35.9 t.
DMCO(B). Bombardier Derby 2004. 12/52. 43.3 t.

Note: 375 801–820 are fitted with de-icing equipment. TSO weighs 36.5 t.

375801	**CN**	E	*SE*	RM	73301	79001	78201	73701
375802	**CN**	E	*SE*	RM	73302	79002	78202	73702
375803	**CN**	E	*SE*	RM	73303	79003	78203	73703
375804	**CN**	E	*SE*	RM	73304	79004	78204	73704

375805	**CN**	E	*SE*	RM	73305	79005	78205	73705
375806	**CN**	E	*SE*	RM	73306	79006	78206	73706
375807	**CN**	E	*SE*	RM	73307	79007	78207	73707
375808	**CN**	E	*SE*	RM	73308	79008	78208	73708
375809	**CN**	E	*SE*	RM	73309	79009	78209	73709
375810	**CN**	E	*SE*	RM	73310	79010	78210	73710
375811	**CN**	E	*SE*	RM	73311	79011	78211	73711
375812	**CN**	E	*SE*	RM	73312	79012	78212	73712
375813	**CN**	E	*SE*	RM	73313	79013	78213	73713
375814	**CN**	E	*SE*	RM	73314	79014	78214	73714
375815	**CN**	E	*SE*	RM	73315	79015	78215	73715
375816	**CN**	E	*SE*	RM	73316	79016	78216	73716
375817	**CN**	E	*SE*	RM	73317	79017	78217	73717
375818	**CN**	E	*SE*	RM	73318	79018	78218	73718
375819	**CN**	E	*SE*	RM	73319	79019	78219	73719
375820	**CN**	E	*SE*	RM	73320	79020	78220	73720
375821	**CN**	E	*SE*	RM	73321	79021	78221	73721
375822	**CN**	E	*SE*	RM	73322	79022	78222	73722
375823	**CN**	E	*SE*	RM	73323	79023	78223	73723
375824	**CN**	E	*SE*	RM	73324	79024	78224	73724
375825	**CN**	E	*SE*	RM	73325	79025	78225	73725
375826	**CN**	E	*SE*	RM	73326	79026	78226	73726
375827	**CN**	E	*SE*	RM	73327	79027	78227	73727
375828	**CN**	E	*SE*	RM	73328	79028	78228	73728
375829	**CN**	E	*SE*	RM	73329	79029	78229	73729
375830	**CN**	E	*SE*	RM	73330	79030	78230	73730

Name (carried on one side of each MSO or TSO):

375830   City of London

Class 375/9. Outer suburban units. 750 V DC only. DMCO–MSO–TSO–DMCO.

DMCO(A). Bombardier Derby 2003–04. 12/59. 43.4 t.
MSO. Bombardier Derby 2003–04. –/73 1T. 39.3 t.
TSO. Bombardier Derby 2003–04. –/59 1TD 2W. 35.6 t.
DMCO(B). Bombardier Derby 2003–04. 12/59. 43.4 t.

375901	**CN**	E	*SE*	RM	73331	79031	79061	73731
375902	**CN**	E	*SE*	RM	73332	79032	79062	73732
375903	**CN**	E	*SE*	RM	73333	79033	79063	73733
375904	**CN**	E	*SE*	RM	73334	79034	79064	73734
375905	**CN**	E	*SE*	RM	73335	79035	79065	73735
375906	**CN**	E	*SE*	RM	73336	79036	79066	73736
375907	**CN**	E	*SE*	RM	73337	79037	79067	73737
375908	**CN**	E	*SE*	RM	73338	79038	79068	73738
375909	**CN**	E	*SE*	RM	73339	79039	79069	73739
375910	**CN**	E	*SE*	RM	73340	79040	79070	73740
375911	**CN**	E	*SE*	RM	73341	79041	79071	73741
375912	**CN**	E	*SE*	RM	73342	79042	79072	73742
375913	**CN**	E	*SE*	RM	73343	79043	79073	73743
375914	**CN**	E	*SE*	RM	73344	79044	79074	73744
375915	**CN**	E	*SE*	RM	73345	79045	79075	73745

375916	**CN**	E	*SE*	RM	73346	79046	79076	73746
375917	**CN**	E	*SE*	RM	73347	79047	79077	73747
375918	**CN**	E	*SE*	RM	73348	79048	79078	73748
375919	**CN**	E	*SE*	RM	73349	79049	79079	73749
375920	**CN**	E	*SE*	RM	73350	79050	79080	73750
375921	**CN**	E	*SE*	RM	73351	79051	79081	73751
375922	**CN**	E	*SE*	RM	73352	79052	79082	73752
375923	**CN**	E	*SE*	RM	73353	79053	79083	73753
375924	**CN**	E	*SE*	RM	73354	79054	79084	73754
375925	**CN**	E	*SE*	RM	73355	79055	79085	73755
375926	**CN**	E	*SE*	RM	73356	79056	79086	73756
375927	**CN**	E	*SE*	RM	73357	79057	79087	73757

CLASS 376   ELECTROSTAR   BOMBARDIER DERBY

Inner suburban units.

Formation: DMSO–MSO–TSO–MSO–DMSO.
System: 750 V DC third rail.
Construction: Welded aluminium alloy underframe, sides and roof with steel ends. All sections bolted together.
Traction Motors: Two Bombardier asynchronous of 250 kW.
Wheel Arrangement: 2-Bo + 2-Bo + 2-2 + Bo-2 + Bo-2.
Braking: Disc & regenerative.   **Dimensions:** 20.40/19.99 x 2.80 m.
Bogies: Bombardier P3-25/T3-25.   **Couplers:** Dellner 12.
Gangways: Within unit.   **Control System:** IGBT Inverter.
Doors: Sliding.   **Maximum Speed:** 75 mph.
Heating & ventilation: Pressure heating and ventilation.
Seating Layout: 2+2 low density facing.
Multiple Working: Within class and with Classes 375, 377, 378 and 379.

DMSO(A). Bombardier Derby 2004–05. –/36(6) 1W. 42.1 t.
MSO. Bombardier Derby 2004–05. –/48. 36.2 t.
TSO. Bombardier Derby 2004–05. –/48. 36.3 t.
DMSO(B). Bombardier Derby 2004–05. –/36(6) 1W. 42.1 t.

376001	**CN**	E	*SE*	SG	61101	63301	64301	63501	61601
376002	**CN**	E	*SE*	SG	61102	63302	64302	63502	61602
376003	**CN**	E	*SE*	SG	61103	63303	64303	63503	61603
376004	**CN**	E	*SE*	SG	61104	63304	64304	63504	61604
376005	**CN**	E	*SE*	SG	61105	63305	64305	63505	61605
376006	**CN**	E	*SE*	SG	61106	63306	64306	63506	61606
376007	**CN**	E	*SE*	SG	61107	63307	64307	63507	61607
376008	**CN**	E	*SE*	SG	61108	63308	64308	63508	61608
376009	**CN**	E	*SE*	SG	61109	63309	64309	63509	61609
376010	**CN**	E	*SE*	SG	61110	63310	64310	63510	61610
376011	**CN**	E	*SE*	SG	61111	63311	64311	63511	61611
376012	**CN**	E	*SE*	SG	61112	63312	64312	63512	61612
376013	**CN**	E	*SE*	SG	61113	63313	64313	63513	61613
376014	**CN**	E	*SE*	SG	61114	63314	64314	63514	61614
376015	**CN**	E	*SE*	SG	61115	63315	64315	63515	61615
376016	**CN**	E	*SE*	SG	61116	63316	64316	63516	61616

376 017	**CN**	E	*SE*	SG	61117 63317 64317 63517 61617
376 018	**CN**	E	*SE*	SG	61118 63318 64318 63518 61618
376 019	**CN**	E	*SE*	SG	61119 63319 64319 63519 61619
376 020	**CN**	E	*SE*	SG	61120 63320 64320 63520 61620
376 021	**CN**	E	*SE*	SG	61121 63321 64321 63521 61621
376 022	**CN**	E	*SE*	SG	61122 63322 64322 63522 61622
376 023	**CN**	E	*SE*	SG	61123 63323 64323 63523 61623
376 024	**CN**	E	*SE*	SG	61124 63324 64324 63524 61624
376 025	**CN**	E	*SE*	SG	61125 63325 64325 63525 61625
376 026	**CN**	E	*SE*	SG	61126 63326 64326 63526 61626
376 027	**CN**	E	*SE*	SG	61127 63327 64327 63527 61627
376 028	**CN**	E	*SE*	SG	61128 63328 64328 63528 61628
376 029	**CN**	E	*SE*	SG	61129 63329 64329 63529 61629
376 030	**CN**	E	*SE*	SG	61130 63330 64330 63530 61630
376 031	**CN**	E	*SE*	SG	61131 63331 64331 63531 61631
376 032	**CN**	E	*SE*	SG	61132 63332 64332 63532 61632
376 033	**CN**	E	*SE*	SG	61133 63333 64333 63533 61633
376 034	**CN**	E	*SE*	SG	61134 63334 64334 63534 61634
376 035	**CN**	E	*SE*	SG	61135 63335 64335 63535 61635
376 036	**CN**	E	*SE*	SG	61136 63336 64336 63536 61636

CLASS 377   ELECTROSTAR   BOMBARDIER DERBY

Express and outer suburban units.

Formations: Various.
Systems: 25 kV AC overhead/750 V DC third rail or third rail only with provision for retro-fitting of AC equipment.
Construction: Welded aluminium alloy underframe, sides and roof with steel ends. All sections bolted together.
Traction Motors: Two Bombardier asynchronous of 250 kW.
Wheel Arrangement: 2-Bo(+ 2-Bo) + 2-2 + Bo-2.
Braking: Disc & regenerative.   **Dimensions:** 20.39/19.99 x 2.80 m.
Bogies: Bombardier P3-25/T3-25.   **Couplers:** Dellner 12.
Gangways: Throughout.   **Control System:** IGBT Inverter.
Doors: Sliding plug.   **Maximum Speed:** 100 mph.
Heating & ventilation: Air conditioning.
Seating Layout: Various.
Multiple Working: Within class and with Classes 375, 376, 378 and 379.

Class 377/1. 750 V DC only. DMCO–MSO–TSO–DMCO.
Seating layout: 1: 2+2 facing/unidirectional, 2: 2+2 facing/unidirectional (377 101–119), 3+2 and 2+2 facing/unidirectional (377 120–164) (3+2 seating in middle cars only 377 140–164).

DMCO(A). Bombardier Derby 2002–03. 12/48 (s 12/56). 44.8 t.
MSO. Bombardier Derby 2002–03. –/62 (s –/70, t –/69). 1T. 39.0 t.
TSO. Bombardier Derby 2002–03. –/52 (s –/60, t –/57). 1TD 2W. 35.4 t.
DMCO(B). Bombardier Derby 2002–03. 12/48 (s 12/56). 43.4 t.

377 101	**SN**	P	*SN*	BI	78501 77101 78901 78701
377 102	**SN**	P	*SN*	BI	78502 77102 78902 78702
377 103	**SN**	P	*SN*	BI	78503 77103 78903 78703

377 104		**SN**	P	*SN*	Bl	78504	77104	78904	78704
377 105		**SN**	P	*SN*	Bl	78505	77105	78905	78705
377 106		**SN**	P	*SN*	Bl	78506	77106	78906	78706
377 107		**SN**	P	*SN*	Bl	78507	77107	78907	78707
377 108		**SN**	P	*SN*	Bl	78508	77108	78908	78708
377 109		**SN**	P	*SN*	Bl	78509	77109	78909	78709
377 110		**SN**	P	*SN*	Bl	78510	77110	78910	78710
377 111		**SN**	P	*SN*	Bl	78511	77111	78911	78711
377 112		**SN**	P	*SN*	Bl	78512	77112	78912	78712
377 113		**SN**	P	*SN*	Bl	78513	77113	78913	78713
377 114		**SN**	P	*SN*	Bl	78514	77114	78914	78714
377 115		**SN**	P	*SN*	Bl	78515	77115	78915	78715
377 116		**SN**	P	*SN*	Bl	78516	77116	78916	78716
377 117		**SN**	P	*SN*	Bl	78517	77117	78917	78717
377 118		**SN**	P	*SN*	Bl	78518	77118	78918	78718
377 119		**SN**	P	*SN*	Bl	78519	77119	78919	78719
377 120	s	**SN**	P	*SN*	Bl	78520	77120	78920	78720
377 121	s	**SN**	P	*SN*	Bl	78521	77121	78921	78721
377 122	s	**SN**	P	*SN*	Bl	78522	77122	78922	78722
377 123	s	**SN**	P	*SN*	Bl	78523	77123	78923	78723
377 124	s	**SN**	P	*SN*	Bl	78524	77124	78924	78724
377 125	s	**SN**	P	*SN*	Bl	78525	77125	78925	78725
377 126	s	**SN**	P	*SN*	Bl	78526	77126	78926	78726
377 127	s	**SN**	P	*SN*	Bl	78527	77127	78927	78727
377 128	s	**SN**	P	*SN*	Bl	78528	77128	78928	78728
377 129	s	**SN**	P	*SN*	Bl	78529	77129	78929	78729
377 130	s	**SN**	P	*SN*	Bl	78530	77130	78930	78730
377 131	s	**SN**	P	*SN*	Bl	78531	77131	78931	78731
377 132	s	**SN**	P	*SN*	Bl	78532	77132	78932	78732
377 133	s	**SN**	P	*SN*	Bl	78533	77133	78933	78733
377 134	s	**SN**	P	*SN*	Bl	78534	77134	78934	78734
377 135	s	**SN**	P	*SN*	Bl	78535	77135	78935	78735
377 136	s	**SN**	P	*SN*	Bl	78536	77136	78936	78736
377 137	s	**SN**	P	*SN*	Bl	78537	77137	78937	78737
377 138	s	**SN**	P	*SN*	Bl	78538	77138	78938	78738
377 139	s	**SN**	P	*SN*	Bl	78539	77139	78939	78739
377 140	t	**SN**	P	*SN*	Bl	78540	77140	78940	78740
377 141	t	**SN**	P	*SN*	Bl	78541	77141	78941	78741
377 142	t	**SN**	P	*SN*	Bl	78542	77142	78942	78742
377 143	t	**SN**	P	*SN*	Bl	78543	77143	78943	78743
377 144	t	**SN**	P	*SN*	Bl	78544	77144	78944	78744
377 145	t	**SN**	P	*SN*	Bl	78545	77145	78945	78745
377 146	t	**SN**	P	*SN*	Bl	78546	77146	78946	78746
377 147	t	**SN**	P	*SN*	Bl	78547	77147	78947	78747
377 148	t	**SN**	P	*SN*	Bl	78548	77148	78948	78748
377 149	t	**SN**	P	*SN*	Bl	78549	77149	78949	78749
377 150	t	**SN**	P	*SN*	Bl	78550	77150	78950	78750
377 151	t	**SN**	P	*SN*	Bl	78551	77151	78951	78751
377 152	t	**SN**	P	*SN*	Bl	78552	77152	78952	78752
377 153	t	**SN**	P	*SN*	Bl	78553	77153	78953	78753
377 154	t	**SN**	P	*SN*	Bl	78554	77154	78954	78754

377 155	t	**SN**	P	*SN*	BI	78555	77155	78955	78755
377 156	t	**SN**	P	*SN*	BI	78556	77156	78956	78756
377 157	t	**SN**	P	*SN*	BI	78557	77157	78957	78757
377 158	t	**SN**	P	*SN*	BI	78558	77158	78958	78758
377 159	t	**SN**	P	*SN*	BI	78559	77159	78959	78759
377 160	t	**SN**	P	*SN*	BI	78560	77160	78960	78760
377 161	t	**SN**	P	*SN*	BI	78561	77161	78961	78761
377 162	t	**SN**	P	*SN*	BI	78562	77162	78962	78762
377 163	t	**SN**	P	*SN*	BI	78563	77163	78963	78763
377 164	t	**SN**	P	*SN*	BI	78564	77164	78964	78764

Class 377/2. 25 kV AC/750 V DC. DMCO–MSO–PTSO–DMCO. These dual-voltage units are used on the South Croydon–Milton Keynes cross-London service. Three units transferred to First Capital Connect in 2011.
Seating layout: 1: 2+2 facing/unidirectional, 2: 2+2 and 3+2 facing/unidirectional (3+2 seating in middle cars only).

DMCO(A). Bombardier Derby 2003–04. 12/48. 44.2 t.
MSO. Bombardier Derby 2003–04. –/69 1T. 39.8 t.
PTSO. Bombardier Derby 2003–04. –/57 1TD 2W. 40.1 t.
DMCO(B). Bombardier Derby 2003–04. 12/48. 44.2 t.

377 201		**SN**	P	*SN*	SU	78571	77171	78971	78771
377 202		**SN**	P	*SN*	SU	78572	77172	78972	78772
377 203		**SN**	P	*SN*	SU	78573	77173	78973	78773
377 204		**SN**	P	*SN*	SU	78574	77174	78974	78774
377 205		**SN**	P	*SN*	SU	78575	77175	78975	78775
377 206		**SN**	P	*SN*	SU	78576	77176	78976	78776
377 207		**FU**	P	*FC*	BF	78577	77177	78977	78777
377 208		**SN**	P	*SN*	SU	78578	77178	78978	78778
377 209		**SN**	P	*SN*	SU	78579	77179	78979	78779
377 210		**SN**	P	*SN*	SU	78580	77180	78980	78780
377 211		**FU**	P	*FC*	BF	78581	77181	78981	78781
377 212		**FU**	P	*FC*	BF	78582	77182	78982	78782
377 213		**SN**	P	*SN*	SU	78583	77183	78983	78783
377 214		**SN**	P	*SN*	SU	78584	77184	78984	78784
377 215		**SN**	P	*SN*	SU	78585	77185	78985	78785

Class 377/3. 750 V DC only. DMCO–TSO–DMCO.
Seating Layout: 1: 2+2 facing/unidirectional, 2: 2+2 facing/unidirectional.

Note: Units built as Class 375, but renumbered in the Class 377/3 range when fitted with Dellner couplers.

DMCO(A). Bombardier Derby 2001–02. 12/48. 43.5 t.
TSO. Bombardier Derby 2001–02. –/56 1TD 2W. 35.4 t.
DMCO(B). Bombardier Derby 2001–02. 12/48. 43.5 t.

377 301	(375 311)	**SN**	P	*SN*	SU	68201	74801	68401
377 302	(375 312)	**SN**	P	*SN*	SU	68202	74802	68402
377 303	(375 313)	**SN**	P	*SN*	SU	68203	74803	68403
377 304	(375 314)	**SN**	P	*SN*	SU	68204	74804	68404
377 305	(375 315)	**SN**	P	*SN*	SU	68205	74805	68405
377 306	(375 316)	**SN**	P	*SN*	SU	68206	74806	68406

377 307	(375 317)	**SN**	P	*SN*	SU	68207	74807	68407
377 308	(375 318)	**SN**	P	*SN*	SU	68208	74808	68408
377 309	(375 319)	**SN**	P	*SN*	SU	68209	74809	68409
377 310	(375 320)	**SN**	P	*SN*	SU	68210	74810	68410
377 311	(375 321)	**SN**	P	*SN*	SU	68211	74811	68411
377 312	(375 322)	**SN**	P	*SN*	SU	68212	74812	68412
377 313	(375 323)	**SN**	P	*SN*	SU	68213	74813	68413
377 314	(375 324)	**SN**	P	*SN*	SU	68214	74814	68414
377 315	(375 325)	**SN**	P	*SN*	SU	68215	74815	68415
377 316	(375 326)	**SN**	P	*SN*	SU	68216	74816	68416
377 317	(375 327)	**SN**	P	*SN*	SU	68217	74817	68417
377 318	(375 328)	**SN**	P	*SN*	SU	68218	74818	68418
377 319	(375 329)	**SN**	P	*SN*	SU	68219	74819	68419
377 320	(375 330)	**SN**	P	*SN*	SU	68220	74820	68420
377 321	(375 331)	**SN**	P	*SN*	SU	68221	74821	68421
377 322	(375 332)	**SN**	P	*SN*	SU	68222	74822	68422
377 323	(375 333)	**SN**	P	*SN*	SU	68223	74823	68423
377 324	(375 334)	**SN**	P	*SN*	SU	68224	74824	68424
377 325	(375 335)	**SN**	P	*SN*	SU	68225	74825	68425
377 326	(375 336)	**SN**	P	*SN*	SU	68226	74826	68426
377 327	(375 337)	**SN**	P	*SN*	SU	68227	74827	68427
377 328	(375 338)	**SN**	P	*SN*	SU	68228	74828	68428

Class 377/4. 750 V DC only. DMCO–MSO–TSO–DMCO.
Seating Layout: 1: 2+2 facing/two seats longitudinal, 2: 2+2 and 3+2 facing/unidirectional (3+2 seating in middle cars only).

DMCO(A). Bombardier Derby 2004–05. 10/48. 43.1 t.
MSO. Bombardier Derby 2004–05. –/69 1T. 39.3 t.
TSO. Bombardier Derby 2004–05. –/56 1TD 2W. 35.3 t.
DMCO(B). Bombardier Derby 2004–05. 10/48. 43.2 t.

377 401	**SN**	P	*SN*	BI	73401	78801	78601	73801
377 402	**SN**	P	*SN*	BI	73402	78802	78602	73802
377 403	**SN**	P	*SN*	BI	73403	78803	78603	73803
377 404	**SN**	P	*SN*	BI	73404	78804	78604	73804
377 405	**SN**	P	*SN*	BI	73405	78805	78605	73805
377 406	**SN**	P	*SN*	BI	73406	78806	78606	73806
377 407	**SN**	P	*SN*	BI	73407	78807	78607	73807
377 408	**SN**	P	*SN*	BI	73408	78808	78608	73808
377 409	**SN**	P	*SN*	BI	73409	78809	78609	73809
377 410	**SN**	P	*SN*	BI	73410	78810	78610	73810
377 411	**SN**	P	*SN*	BI	73411	78811	78611	73811
377 412	**SN**	P	*SN*	BI	73412	78812	78612	73812
377 413	**SN**	P	*SN*	BI	73413	78813	78613	73813
377 414	**SN**	P	*SN*	BI	73414	78814	78614	73814
377 415	**SN**	P	*SN*	BI	73415	78815	78615	73815
377 416	**SN**	P	*SN*	BI	73416	78816	78616	73816
377 417	**SN**	P	*SN*	BI	73417	78817	78617	73817
377 418	**SN**	P	*SN*	BI	73418	78818	78618	73818
377 419	**SN**	P	*SN*	BI	73419	78819	78619	73819
377 420	**SN**	P	*SN*	BI	73420	78820	78620	73820

377421	**SN**	P	*SN*	BI	73421	78821	78621	73821
377422	**SN**	P	*SN*	BI	73422	78822	78622	73822
377423	**SN**	P	*SN*	BI	73423	78823	78623	73823
377424	**SN**	P	*SN*	BI	73424	78824	78624	73824
377425	**SN**	P	*SN*	BI	73425	78825	78625	73825
377426	**SN**	P	*SN*	BI	73426	78826	78626	73826
377427	**SN**	P	*SN*	BI	73427	78827	78627	73827
377428	**SN**	P	*SN*	BI	73428	78828	78628	73828
377429	**SN**	P	*SN*	BI	73429	78829	78629	73829
377430	**SN**	P	*SN*	BI	73430	78830	78630	73830
377431	**SN**	P	*SN*	BI	73431	78831	78631	73831
377432	**SN**	P	*SN*	BI	73432	78832	78632	73832
377433	**SN**	P	*SN*	BI	73433	78833	78633	73833
377434	**SN**	P	*SN*	BI	73434	78834	78634	73834
377435	**SN**	P	*SN*	BI	73435	78835	78635	73835
377436	**SN**	P	*SN*	BI	73436	78836	78636	73836
377437	**SN**	P	*SN*	BI	73437	78837	78637	73837
377438	**SN**	P	*SN*	BI	73438	78838	78638	73838
377439	**SN**	P	*SN*	BI	73439	78839	78639	73839
377440	**SN**	P	*SN*	BI	73440	78840	78640	73840
377441	**SN**	P	*SN*	BI	73441	78841	78641	73841
377442	**SN**	P	*SN*	BI	73442	78842	78642	73842
377443	**SN**	P	*SN*	BI	73443	78843	78643	73843
377444	**SN**	P	*SN*	BI	73444	78844	78644	73844
377445	**SN**	P	*SN*	BI	73445	78845	78645	73845
377446	**SN**	P	*SN*	BI	73446	78846	78646	73846
377447	**SN**	P	*SN*	BI	73447	78847	78647	73847
377448	**SN**	P	*SN*	BI	73448	78848	78648	73848
377449	**SN**	P	*SN*	BI	73449	78849	78649	73849
377450	**SN**	P	*SN*	BI	73450	78850	78650	73850
377451	**SN**	P	*SN*	BI	73451	78851	78651	73851
377452	**SN**	P	*SN*	BI	73452	78852	78652	73852
377453	**SN**	P	*SN*	BI	73453	78853	78653	73853
377454	**SN**	P	*SN*	BI	73454	78854	78654	73854
377455	**SN**	P	*SN*	BI	73455	78855	78655	73855
377456	**SN**	P	*SN*	BI	73456	78856	78656	73856
377457	**SN**	P	*SN*	BI	73457	78857	78657	73857
377458	**SN**	P	*SN*	BI	73458	78858	78658	73858
377459	**SN**	P	*SN*	BI	73459	78859	78659	73859
377460	**SN**	P	*SN*	BI	73460	78860	78660	73860
377461	**SN**	P	*SN*	BI	73461	78861	78661	73861
377462	**SN**	P	*SN*	BI	73462	78862	78662	73862
377463	**SN**	P	*SN*	BI	73463	78863	78663	73863
377464	**SN**	P	*SN*	BI	73464	78864	78664	73864
377465	**SN**	P	*SN*	BI	73465	78865	78665	73865
377466	**SN**	P	*SN*	BI	73466	78866	78666	73866
377467	**SN**	P	*SN*	BI	73467	78867	78667	73867
377468	**SN**	P	*SN*	BI	73468	78868	78668	73868
377469	**SN**	P	*SN*	BI	73469	78869	78669	73869
377470	**SN**	P	*SN*	BI	73470	78870	78670	73870
377471	**SN**	P	*SN*	BI	73471	78871	78671	73871

377472	**SN**	P	*SN*	Bl	73472	78872	78672	73872
377473	**SN**	P	*SN*	Bl	73473	78873	78673	73873
377474	**SN**	P	*SN*	Bl	73474	78874	78674	73874
377475	**SN**	P	*SN*	Bl	73475	78875	78675	73875

Class 377/5. 25 kV AC/750 V DC. DMCO–MSO–PTSO–DMCO. Dual voltage First Capital Connect units (sub-leased from Southern). Details as Class 377/2 unless stated.

DMCO(A). Bombardier Derby 2008–09. 10/48. 43.1 t.
MSO. Bombardier Derby 2008–09. –/69 1T. 40.3 t.
PTSO. Bombardier Derby 2008–09. –/56 1TD 2W. 40.6 t.
DMCO(B). Bombardier Derby 2008–09. 10/48. 43.1 t.

377501	**FU**	P	*FC*	BF	73501	75901	74901	73601
377502	**FU**	P	*FC*	BF	73502	75902	74902	73602
377503	**FU**	P	*FC*	BF	73503	75903	74903	73603
377504	**FU**	P	*FC*	BF	73504	75904	74904	73604
377505	**FU**	P	*FC*	BF	73505	75905	74905	73605
377506	**FU**	P	*FC*	BF	73506	75906	74906	73606
377507	**FU**	P	*FC*	BF	73507	75907	74907	73607
377508	**FU**	P	*FC*	BF	73508	75908	74908	73608
377509	**FU**	P	*FC*	BF	73509	75909	74909	73609
377510	**FU**	P	*FC*	BF	73510	75910	74910	73610
377511	**FU**	P	*FC*	BF	73511	75911	74911	73611
377512	**FU**	P	*FC*	BF	73512	75912	74912	73612
377513	**FU**	P	*FC*	BF	73513	75913	74913	73613
377514	**FU**	P	*FC*	BF	73514	75914	74914	73614
377515	**FU**	P	*FC*	BF	73515	75915	74915	73615
377516	**FU**	P	*FC*	BF	73516	75916	74916	73616
377517	**FU**	P	*FC*	BF	73517	75917	74917	73617
377518	**FU**	P	*FC*	BF	73518	75918	74918	73618
377519	**FU**	P	*FC*	BF	73519	75919	74919	73619
377520	**FU**	P	*FC*	BF	73520	75920	74920	73620
377521	**FU**	P	*FC*	BF	73521	75921	74921	73621
377522	**FU**	P	*FC*	BF	73522	75922	74922	73622
377523	**FU**	P	*FC*	BF	73523	75923	74923	73623

Class 377/6. 750 V DC. DMSO–MSO–TSO–MSO–DMCO. On order for Southern metro services and due for delivery 2013. Full details awaited; numbering series is provisional.

DMSO. Bombardier Derby 2012–13.   .   t.
MSO. Bombardier Derby 2012–13.   .   t.
TSO. Bombardier Derby 2012–13.   .   t.
MSO. Bombardier Derby 2012–13.   .   t.
DMCO. Bombardier Derby 2012–13.   .   t.

377601	P	70101	70201	70301	70401	70501
377602	P	70102	70202	70302	70402	70502
377603	P	70103	70203	70303	70403	70503
377604	P	70104	70204	70304	70404	70504
377605	P	70105	70205	70305	70405	70505
377606	P	70106	70206	70306	70406	70506

377607	P	70107	70207	70307	70407	70507
377608	P	70108	70208	70308	70408	70508
377609	P	70109	70209	70309	70409	70509
377610	P	70110	70210	70310	70410	70510
377611	P	70111	70211	70311	70411	70511
377612	P	70112	70212	70312	70412	70512
377613	P	70113	70213	70313	70413	70513
377614	P	70114	70214	70314	70414	70514
377615	P	70115	70215	70315	70415	70515
377616	P	70116	70216	70316	70416	70516
377617	P	70117	70217	70317	70417	70517
377618	P	70118	70218	70318	70418	70518
377619	P	70119	70219	70319	70419	70519
377620	P	70120	70220	70320	70420	70520
377621	P	70121	70221	70321	70421	70521
377622	P	70122	70222	70322	70422	70522
377623	P	70123	70223	70323	70423	70523
377624	P	70124	70224	70324	70424	70524
377625	P	70125	70225	70325	70425	70525
377626	P	70126	70226	70326	70426	70526

CLASS 378   CAPITALSTAR   BOMBARDIER DERBY

57 new Class 378 suburban Electrostars (designated Capitalstars by TfL) used by London Overground.

Formation: DMSO–MSO–TSO–DMSO or DMSO–MSO–PTSO–DMSO.
System: Class 378/1 750 V DC third rail only. Class 378/2 25 kV AC overhead and 750 V DC third rail.
Construction: Welded aluminium alloy underframe, sides and roof with steel ends. All sections bolted together.
Traction Motors: Two Bombardier asynchronous of 250 kW.
Wheel Arrangement: 2-Bo + 2-Bo + 2-2 + Bo-2.
Braking: Disc & regenerative. **Dimensions:** 20.46/20.14 x 2.80 m.
Bogies: Bombardier P3-25/T3-25. **Couplers:** Dellner 12.
Gangways: Within unit + end doors. **Control System:** IGBT Inverter.
Doors: Sliding. **Maximum Speed:** 75 mph.
Heating & ventilation: Air conditioning.
Seating Layout: Longitudinal ("tube style") low density.
Multiple Working: Within class and with Classes 375, 376, 377 and 379.

Class 378/1. 750 V DC. DMSO–MSO–TSO–DMSO. Third rail only units used on East London Line services. Provision for retro-fitting as dual voltage.

DMSO(A). Bombardier Derby 2009–10. –/36. 43.5 t.
MSO. Bombardier Derby 2009–10. –/40. 39.4 t.
TSO. Bombardier Derby 2009–10. –/34(6) 2W. 34.3t.
DMSO(B). Bombardier Derby 2009–10. –/36. 43.1 t.

Note: 378 150–154 are fitted with de-icing equipment. TSO weighs 34.8t.

378 135	**LO**	QW *LO*	NG	38035	38235	38335	38135	
378 136	**LO**	QW *LO*	NG	38036	38236	38336	38136	

378137	LO	QW	LO	NG	38037	38237	38337	38137
378138	LO	QW	LO	NG	38038	38238	38338	38138
378139	LO	QW	LO	NG	38039	38239	38339	38139
378140	LO	QW	LO	NG	38040	38240	38340	38140
378141	LO	QW	LO	NG	38041	38241	38341	38141
378142	LO	QW	LO	NG	38042	38242	38342	38142
378143	LO	QW	LO	NG	38043	38243	38343	38143
378144	LO	QW	LO	NG	38044	38244	38344	38144
378145	LO	QW	LO	NG	38045	38245	38345	38145
378146	LO	QW	LO	NG	38046	38246	38346	38146
378147	LO	QW	LO	NG	38047	38247	38347	38147
378148	LO	QW	LO	NG	38048	38248	38348	38148
378149	LO	QW	LO	NG	38049	38249	38349	38149
378150	LO	QW	LO	NG	38050	38250	38350	38150
378151	LO	QW	LO	NG	38051	38251	38351	38151
378152	LO	QW	LO	NG	38052	38252	38352	38152
378153	LO	QW	LO	NG	38053	38253	38353	38153
378154	LO	QW	LO	NG	38054	38254	38354	38154

Class 378/2. 25 kV AC/750 V DC. DMSO–MSO–PTSO–DMSO. Dual voltage units mainly used on North London Railway services. 378201–224 built as 378001–024 (3-car units) and extended to 4-car units in 2010.

DMSO(A). Bombardier Derby 2008–11. –/36. 43.2 t.
MSO. Bombardier Derby 2008–11. –/40. 39.8 t.
PTSO. Bombardier Derby 2008–11. –/34(6) 2W. 39.0 t.
DMSO(B). Bombardier Derby 2008–11. –/36. 42.8 t.

Advertising livery: 378 211 and 378 221 Lycamobile (white).

Note: 378 216–220 are fitted with de-icing equipment.

378201	LO	QW	LO	NG	38001	38201	38301	38101
378202	LO	QW	LO	NG	38002	38202	38302	38102
378203	LO	QW	LO	NG	38003	38203	38303	38103
378204	LO	QW	LO	NG	38004	38204	38304	38104
378205	LO	QW	LO	NG	38005	38205	38305	38105
378206	LO	QW	LO	NG	38006	38206	38306	38106
378207	LO	QW	LO	NG	38007	38207	38307	38107
378208	LO	QW	LO	NG	38008	38208	38308	38108
378209	LO	QW	LO	NG	38009	38209	38309	38109
378210	LO	QW	LO	NG	38010	38210	38310	38110
378211	AL	QW	LO	NG	38011	38211	38311	38111
378212	LO	QW	LO	NG	38012	38212	38312	38112
378213	LO	QW	LO	NG	38013	38213	38313	38113
378214	LO	QW	LO	NG	38014	38214	38314	38114
378215	LO	QW	LO	NG	38015	38215	38315	38115
378216	LO	QW	LO	NG	38016	38216	38316	38116
378217	LO	QW	LO	NG	38017	38217	38317	38117
378218	LO	QW	LO	NG	38018	38218	38318	38118
378219	LO	QW	LO	NG	38019	38219	38319	38119
378220	LO	QW	LO	NG	38020	38220	38320	38120
378221	AL	QW	LO	NG	38021	38221	38321	38121

378 222	LO	QW	LO	NG	38022	38222	38322	38122
378 223	LO	QW	LO	NG	38023	38223	38323	38123
378 224	LO	QW	LO	NG	38024	38224	38324	38124
378 225	LO	QW	LO	NG	38025	38225	38325	38125
378 226	LO	QW	LO	NG	38026	38226	38326	38126
378 227	LO	QW	LO	NG	38027	38227	38327	38127
378 228	LO	QW	LO	NG	38028	38228	38328	38128
378 229	LO	QW	LO	NG	38029	38229	38329	38129
378 230	LO	QW	LO	NG	38030	38230	38330	38130
378 231	LO	QW	LO	NG	38031	38231	38331	38131
378 232	LO	QW	LO	NG	38032	38232	38332	38132
378 233	LO	QW	LO	NG	38033	38233	38333	38133
378 234	LO	QW	LO	NG	38034	38234	38334	38134
378 255	LO	QW	LO	NG	38055	38255	38355	38155
378 256	LO	QW	LO	NG	38056	38256	38356	38156
378 257	LO	QW	LO	NG	38057	38257	38357	38157

Name (carried on 38033):

378 233   Ian Brown CBE

CLASS 379   ELECTROSTAR   BOMBARDIER DERBY

New EMUs used on Liverpool Street–Stansted Airport and Liverpool Street–Cambridge services.

Formation: DMSO–MSO–PTSO–DMCO.
System: 25 kV AC overhead.
Construction: Welded aluminium alloy underframe, sides and roof with steel ends. All sections bolted together.
Traction Motors: Two Bombardier asynchronous of 250 kW.
Wheel Arrangement: 2-Bo + 2-Bo + 2-2 + Bo-2.
Braking: Disc & regenerative.   **Dimensions:** 20.0 x 2.80 m.
Bogies: Bombardier P3-25/T3-25.   **Couplers:** Dellner 12.
Gangways: Throughout.   **Control System:** IGBT Inverter.
Doors: Sliding plug.   **Maximum Speed:** 100 mph.
Heating & ventilation: Air conditioning.
Seating Layout: 1: 2+1 facing. 2: 2+2 facing/unidirectional.
Multiple Working: Within class and with Classes 375, 376, 377 and 378.

DMSO. Bombardier Derby 2010–11. –/60. 42.1 t.
MSO. Bombardier Derby 2010–11. –/62 1T. 38.6 t.
PTSO. Bombardier Derby 2010–11. –/43(2) 1TD 2W. 40.9 t.
DMCO. Bombardier Derby 2010–11. 20/24. 42.3 t.

379 001	NC	LY	EA	IL	61201	61701	61901	62101
379 002	NC	LY	EA	IL	61202	61702	61902	62102
379 003	NC	LY	EA	IL	61203	61703	61903	62103
379 004	NC	LY	EA	IL	61204	61704	61904	62104
379 005	NC	LY	EA	IL	61205	61705	61905	62105
379 006	NC	LY	EA	IL	61206	61706	61906	62106
379 007	NC	LY	EA	IL	61207	61707	61907	62107
379 008	NC	LY	EA	IL	61208	61708	61908	62108

379009	**NC**	LY	*EA*	IL	61209	61709	61909	62109
379010	**NC**	LY	*EA*	IL	61210	61710	61910	62110
379011	**NC**	LY	*EA*	IL	61211	61711	61911	62111
379012	**NC**	LY	*EA*	IL	61212	61712	61912	62112
379013	**NC**	LY	*EA*	IL	61213	61713	61913	62113
379014	**NC**	LY	*EA*	IL	61214	61714	61914	62114
379015	**NC**	LY	*EA*	IL	61215	61715	61915	62115
379016	**NC**	LY	*EA*	IL	61216	61716	61916	62116
379017	**NC**	LY	*EA*	IL	61217	61717	61917	62117
379018	**NC**	LY	*EA*	IL	61218	61718	61918	62118
379019	**NC**	LY	*EA*	IL	61219	61719	61919	62119
379020	**NC**	LY	*EA*	IL	61220	61720	61920	62120
379021	**NC**	LY	*EA*	IL	61221	61721	61921	62121
379022	**NC**	LY	*EA*	IL	61222	61722	61922	62122
379023	**NC**	LY	*EA*	IL	61223	61723	61923	62123
379024	**NC**	LY	*EA*	IL	61224	61724	61924	62124
379025	**NC**	LY	*EA*	IL	61225	61725	61925	62125
379026	**NC**	LY	*EA*	IL	61226	61726	61926	62126
379027	**NC**	LY	*EA*	IL	61227	61727	61927	62127
379028	**NC**	LY	*EA*	IL	61228	61728	61928	62128
379029	**NC**	LY	*EA*	IL	61229	61729	61929	62129
379030	**NC**	LY	*EA*	IL	61230	61730	61930	62130

Names (carried on end cars):

379005 Stansted Express
379011 Ely Cathedral
379012 The West Anglian

379015 City of Cambridge
379025 Go Discover

CLASS 380   DESIRO UK   SIEMENS

New EMUs mainly used on Strathclyde area services from Glasgow Central as well as Edinburgh–North Berwick trains.

Formation: DMSO–PTSO–DMSO or DMSO–PTSO–TSO–DMSO.
System: 25 kV AC overhead.
Construction: Welded aluminium with steel ends.
Traction Motors: Four Siemens 1TB2016-0GB02 asynchronous of 250 kW.
Wheel Arrangement: Bo-Bo + 2-2 (+2-2) + Bo-Bo
Braking: Disc & regenerative.   **Dimensions:** 23.78/23.57 x 2.80 m.
Bogies: SGP SF5000.   **Couplers:** Voith.
Gangways: Throughout.   **Control System:** IGBT Inverter.
Doors: Sliding plug.   **Maximum Speed:** 100 mph.
Heating & ventilation: Air conditioning.
Seating Layout: 2+2 facing/unidirectional.
Multiple Working: Within class.

DMSO(A). Siemens Krefeld 2009–10. –/70. 45.0t.
PTSO. Siemens Krefeld 2009–10. –/57(12) 1TD 2W. 42.7t.
TSO. Siemens Krefeld 2009–10. –/74 1T. 34.8t.
DMSO(B). Siemens Krefeld 2009–10. –/64(5). 44.9t.

Class 380/0. 3-car units.

380001	**SR**	E	*SR*	GW	38501	38601		38701
380002	**SR**	E	*SR*	GW	38502	38602		38702
380003	**SR**	E	*SR*	GW	38503	38603		38703
380004	**SR**	E	*SR*	GW	38504	38604		38704
380005	**SR**	E	*SR*	GW	38505	38605		38705
380006	**SR**	E	*SR*	GW	38506	38606		38706
380007	**SR**	E	*SR*	GW	38507	38607		38707
380008	**SR**	E	*SR*	GW	38508	38608		38708
380009	**SR**	E	*SR*	GW	38509	38609		38709
380010	**SR**	E	*SR*	GW	38510	38610		38710
380011	**SR**	E	*SR*	GW	38511	38611		38711
380012	**SR**	E	*SR*	GW	38512	38612		38712
380013	**SR**	E	*SR*	GW	38513	38613		38713
380014	**SR**	E	*SR*	GW	38514	38614		38714
380015	**SR**	E	*SR*	GW	38515	38615		38715
380016	**SR**	E	*SR*	GW	38516	38616		38716
380017	**SR**	E	*SR*	GW	38517	38617		38717
380018	**SR**	E	*SR*	GW	38518	38618		38718
380019	**SR**	E	*SR*	GW	38519	38619		38719
380020	**SR**	E	*SR*	GW	38520	38620		38720
380021	**SR**	E	*SR*	GW	38521	38621		38721
380022	**SR**	E	*SR*	GW	38522	38622		38722

Class 380/1. 4-car units.

Note: 380 101–108 are fitted with NRN radios for working to North Berwick.

380101	**SR**	E	*SR*	GW	38551	38651	38851	38751
380102	**SR**	E	*SR*	GW	38552	38652	38852	38752
380103	**SR**	E	*SR*	GW	38553	38653	38853	38753
380104	**SR**	E	*SR*	GW	38554	38654	38854	38754
380105	**SR**	E	*SR*	GW	38555	38655	38855	38755
380106	**SR**	E	*SR*	GW	38556	38656	38856	38756
380107	**SR**	E	*SR*	GW	38557	38657	38857	38757
380108	**SR**	E	*SR*	GW	38558	38658	38858	38758
380109	**SR**	E	*SR*	GW	38559	38659	38859	38759
380110	**SR**	E	*SR*	GW	38560	38660	38860	38760
380111	**SR**	E	*SR*	GW	38561	38661	38861	38761
380112	**SR**	E	*SR*	GW	38562	38662	38862	38762
380113	**SR**	E	*SR*	GW	38563	38663	38863	38763
380114	**SR**	E	*SR*	GW	38564	38664	38864	38764
380115	**SR**	E	*SR*	GW	38565	38665	38865	38765
380116	**SR**	E	*SR*	GW	38566	38666	38866	38766

CLASS 390   PENDOLINO   ALSTOM

Tilting West Coast Main Line units.

Formation: DMRFO–MFO–PTFO–MFO–(TSO)–(MSO)–TSO–MSO–PTSRMB–MSO–DMSO.
Construction: Welded aluminium alloy.
Traction Motors: Two Alstom ONIX 800 of 425 kW.
Wheel Arrangement: 1A-A1 + 1A-A1 + 2-2 + 1A-A1 (+ 2-2 + 1A-A1) + 2-2 + 1A-A1 + 2-2 + 1A-A1 + 1A-A1.
Braking: Disc, rheostatic & regenerative.

Dimensions: 24.80/23.90 x 2.73 m.	**Couplers:** Dellner 12.
Bogies: Fiat-SIG.	**Control System:** IGBT Inverter.
Gangways: Within unit.	**Maximum Speed:** 125 mph.
Doors: Sliding plug.	**Heating & ventilation:** Air conditioning.

Seating Layout: 1: 2+1 facing/unidirectional, 2: 2+2 facing/unidirectional.
Multiple Working: Within class. Can also be controlled from Class 57/3 locos.

DMRFO: Alstom Birmingham 2001–05. 18/–. 55.6 t.
MFO(A): Alstom Birmingham 2001–05. 37/–(2) 1TD 1W. 52.0 t.
PTFO: Alstom Birmingham 2001–05. 44/– 1T. 50.1 t.
MFO(B): Alstom Birmingham 2001–05. 46/– 1T. 51.8 t.
(TSO: Alstom Savigliano 2010–12. –/74 1T. 49.2 t.)
(MSO: Alstom Savigliano 2010–12. –/76 1T. 52.2 t.)
TSO: Alstom Birmingham 2001–05. –/76 1T. 45.5 t.
MSO(A): Alstom Birmingham 2001–05. –/62(4) 1TD 1W. 50.0 t.
PTSRMB: Alstom Birmingham 2001–05. –/48. 52.0 t.
MSO(B): Alstom Birmingham 2001–05. –/62(2) 1TD 1W. 51.7 t.
DMSO: Alstom Birmingham 2001–05. –/46 1T. 51.0 t.

Advertising livery: 390004 – Black Alstom vinyls on Virgin silver livery.

Notes: Units up to 390034 were delivered as 8-car sets, without the TSO (688xx). During 2004–05 these units had their 9th cars added.

62 extra vehicles were built 2010–12 to lengthen 31 sets to 11-cars: at the time of writing most of these had been lengthened (the programme was due for completion in December 2012). On renumbering units are being renumbered by adding 100 to the set number. Four new complete 11-car units were also delivered in 2011–12. The final fleet will therefore be 35 11-car sets and 21 9-car sets.

390033 was written off following accident damage in the Lambrigg accident of February 2007.

390 001	**VT**	A	*VW* MA	69101 69401 69501 69601 68801					
				69701 69801 69901 69201					
390 002	**VT**	A	*VW* MA	69102 69402 69502 69602 68802					
				69702 69802 69902 69202					
390 103	**VT**	A	*VW* MA	69103 69403 69503 69603 65303 68903					
				68803 69703 69803 69903 69203					
390 004	**AL**	A	*VW* MA	69104 69404 69504 69604 68804					
				69704 69804 69904 69204					

390005	**VT**	A	VW MA	69105	69405	69505	69605	68805		
				69705	69805	69905	69205			
390006	**VT**	A	VW MA	69106	69406	69506	69606	68806		
				69706	69806	69906	69206			
390107	**VT**	A	VW MA	69107	69407	69507	69607	65307	68907	
				68807	69707	69807	69907	69207		
390008	**VT**	A	VW MA	69108	69408	69508	69608	68808		
				69708	69808	69908	69208			
390009	**VT**	A	VW MA	69109	69409	69509	69609	68809		
				69709	69809	69909	69209			
390010	**VT**	A	VW MA	69110	69410	69510	69610	68810		
				69710	69810	69910	69210			
390011	**VT**	A	VW MA	69111	69411	69511	69611	68811		
				69711	69811	69911	69211			
390112	**VT**	A	VW MA	69112	69412	69512	69612	65312	68912	
				68812	69712	69812	69912	69212		
390013	**VT**	A	VW MA	69113	69413	69513	69613	68813		
				69713	69813	69913	69213			
390114	**VT**	A	VW MA	69114	69414	69514	69614	65314	68914	
				68814	69714	69814	69914	69214		
390015	**VT**	A	VW MA	69115	69415	69515	69615	68815		
				69715	69815	69915	69215			
390016	**VT**	A	VW MA	69116	69416	69516	69616	68816		
				69716	69816	69916	69216			
390117	**VT**	A	VW MA	69117	69417	69517	69617	65317	68917	
				68817	69717	69817	69917	69217		
390118	**VT**	A	VW MA	69118	69418	69518	69618	65318	68918	
				68818	69718	69818	69918	69218		
390119	**VT**	A	VW MA	69119	69419	69519	69619	65319	68919	
				68819	69719	69819	69919	69219		
390020	**VT**	A	VW MA	69120	69420	69520	69620	68820		
				69720	69820	69920	69220			
390121	**VT**	A	VW MA	69121	69421	69521	69621	65321	68921	
				68821	69721	69821	69921	69221		
390122	**VT**	A	VW MA	69122	69422	69522	69622	65322	68922	
				68822	69722	69822	69922	69222		
390023	**VT**	A	VW MA	69123	69423	69523	69623	68823		
				69723	69823	69923	69223			
390124	**VT**	A	VW MA	69124	69424	69524	69624	65324	68924	
				68824	69724	69824	69924	69224		
390125	**VT**	A	VW MA	69125	69425	69525	69625	65325	68925	
				68825	69725	69825	69925	69225		
390126	**VT**	A	VW MA	69126	69426	69526	69626	65326	68926	
				68826	69726	69826	69926	69226		
390127	**VT**	A	VW MA	69127	69427	69527	69627	65327	68927	
				68827	69727	69827	69927	69227		
390128	**VT**	A	VW MA	69128	69428	69528	69628	65328	68928	
				68828	69728	69828	69928	69228		
390129	**VT**	A	VW MA	69129	69429	69529	69629	65329	68929	
				68829	69729	69829	69929	69229		

390 130	**VT**	A	*VW* MA	69130	69430	69530	69630	65330	68930	
				68830	69730	69830	69930	69230		
390 131	**VT**	A	*VW* MA	69131	69431	69531	69631	65331	68931	
				68831	69731	69831	69931	69231		
390 132	**VT**	A	*VW* MA	69132	69432	69532	69632	65332	68932	
				68832	69732	69832	69932	69232		
390 134	**VT**	A	*VW* MA	69134	69434	69534	69634	65334	68934	
				68834	69734	69834	69934	69234		
390 035	**VT**	A	*VW* MA	69135	69435	69535	69635	68835		
				69735	69835	69935	69235			
390 136	**VT**	A	*VW* MA	69136	69436	69536	69636	65336	68936	
				68836	69736	69836	69936	69236		
390 137	**VT**	A	*VW* MA	69137	69437	69537	69637	65337	68937	
				68837	69737	69837	69937	69237		
390 038	**VT**	A	*VW* MA	69138	69438	69538	69638	68838		
				69738	69838	69938	69238			
390 039	**VT**	A	*VW* MA	69139	69439	69539	69639	68839		
				69739	69839	69939	69239			
390 040	**VT**	A	*VW* MA	69140	69440	69540	69640	68840		
				69740	69840	69940	69240			
390 141	**VT**	A	*VW* MA	69141	69441	69541	69641	65341	68941	
				68841	69741	69841	69941	69241		
390 042	**VT**	A	*VW* MA	69142	69442	69542	69642	68842		
				69742	69842	69942	69242			
390 043	**VT**	A	*VW* MA	69143	69443	69543	69643	68843		
				69743	69843	69943	69243			
390 044	**VT**	A	*VW* MA	69144	69444	69544	69644	68844		
				69744	69844	69944	69244			
390 045	**VT**	A	*VW* MA	69145	69445	69545	69645	68845		
				69745	69845	69945	69245			
390 046	**VT**	A	*VW* MA	69146	69446	69546	69646	68846		
				69746	69846	69946	69246			
390 047	**VT**	A	*VW* MA	69147	69447	69547	69647	68847		
				69747	69847	69947	69247			
390 148	**VT**	A	*VW* MA	69148	69448	69548	69648	65348	68948	
				68848	69748	69848	69948	69248		
390 049	**VT**	A	*VW* MA	69149	69449	69549	69649	68849		
				69749	69849	69949	69249			
390 050	**VT**	A	*VW* MA	69150	69450	69550	69650	68850		
				69750	69850	69950	69250			
390 151	**VT**	A	*VW* MA	69151	69451	69551	69651	65351	68951	
				68851	69751	69851	69951	69251		
390 052	**VT**	A	*VW* MA	69152	69452	69552	69652	68852		
				69752	69852	69952	69252			
390 153	**VT**	A	*VW* MA	69153	69453	69553	69653	65353	68953	
				68853	69753	69853	69953	69253		
390 154	**VT**	A	*VW* MA	69154	69454	69554	69654	65354	68954	
				68854	69754	69854	69954	69254		
390 155	**VT**	A	*VW* MA	69155	69455	69555	69655	65355	68955	
				68855	69755	69855	69955	69255		

390156	**VT**	A	*VW* MA	69156	69456	69556	69656	65356	68956
				68856	69756	69856	69956	69256	
390157	**VT**	A	*VW* MA	69157	69457	69557	69657	65357	68957
				68857	69757	69857	69957	69257	

Names (carried on MFO No. 696xx):

390001 Virgin Pioneer	390128 City of Preston
390002 Virgin Angel	390129 City of Stoke-on-Trent
390103 Virgin Hero	390130 City of Edinburgh
390004 Alstom Pendolino	390131 City of Liverpool
390005 City of Wolverhampton	390132 City of Birmingham
390006 Tate Liverpool	390134 City of Carlisle
390107 Virgin Lady	390035 City of Lancaster
390008 Virgin King	390136 City of Coventry
390009 Treaty of Union	390137 Virgin Difference
390010 A Decade of Progress	390038 City of London
390011 City of Lichfield	390039 Virgin Quest
390112 Virgin Star	390040 Virgin Pathfinder
390013 Virgin Spirit	390141 City of Chester
390114 City of Manchester	390042 City of Bangor/Dinas Bangor
390015 Virgin Crusader	390043 Virgin Explorer
390016 Virgin Champion	390044 Virgin Lionheart
390117 Virgin Prince	390045 101 Squadron
390118 Virgin Princess	390046 Virgin Soldiers
390119 Virgin Warrior	390047 CLIC Sargent
390020 Virgin Cavalier	390148 Virgin Harrier
390121 Virgin Dream	390049 Virgin Express
390122 Penny the Pendolino	390050 Virgin Invader
390023 Virgin Glory	390151 Virgin Ambassador
390124 Virgin Venturer	390052 Virgin Knight
390125 Virgin Stagecoach	390153 Mission Accomplished
390126 Virgin Enterprise	390157 Chad Varah
390127 Virgin Buccaneer	

CLASS 395   HS1 DOMESTIC SETS   HITACHI JAPAN

New 6-car dual-voltage units used on Southeastern High Speed services from St Pancras International to Ashford/Dover/Margate via Ramsgate and Faversham.

Formation: PDTSO–MSO–MSO–MSO–MSO–PDTSO.
Systems: 25 kV AC overhead/750 V DC third rail.
Construction: Aluminium.
Traction Motors: Hitachi asynchronous of 210kW.
Wheel Arrangement: 2-2 + Bo-Bo + Bo-Bo + Bo-Bo + Bo-Bo + 2-2.
Braking: Disc, rheostatic & regenerative braking.
Dimensions: 20.88/20.0 x 2.81 m.   **Couplers:** Scharfenberg.
Bogies: Hitachi.   **Control System:** IGBT Inverter.
Gangways: Within unit.   **Maximum Speed:** 140 mph.

Doors: Single-leaf sliding.   **Multiple Working:** Within class only.
Heating & ventilation: Air conditioning.
Seating Layout: 2+2 facing/unidirectional (mainly unidirectional).

PDTSO(A): Hitachi Kasado, Japan 2006–09. –/28(12) 1TD 2W. 46.7 t.
MSO: Hitachi Kasado, Japan 2006–09. –/66. 45.0t–45.7 t.
PDTSO(B): Hitachi Kasado, Japan 2006–09. –/48 1T. 46.7 t.

395001	**SB**	E	*SE*	AD	39011	39012	39013	39014	39015	39016
395002	**SB**	E	*SE*	AD	39021	39022	39023	39024	39025	39026
395003	**SB**	E	*SE*	AD	39031	39032	39033	39034	39035	39036
395004	**SB**	E	*SE*	AD	39041	39042	39043	39044	39045	39046
395005	**SB**	E	*SE*	AD	39051	39052	39053	39054	39055	39056
395006	**SB**	E	*SE*	AD	39061	39062	39063	39064	39065	39066
395007	**SB**	E	*SE*	AD	39071	39072	39073	39074	39075	39076
395008	**SB**	E	*SE*	AD	39081	39082	39083	39084	39085	39086
395009	**SB**	E	*SE*	AD	39091	39092	39093	39094	39095	39096
395010	**SB**	E	*SE*	AD	39101	39102	39103	39104	39105	39106
395011	**SB**	E	*SE*	AD	39111	39112	39113	39114	39115	39116
395012	**SB**	E	*SE*	AD	39121	39122	39123	39124	39125	39126
395013	**SB**	E	*SE*	AD	39131	39132	39133	39134	39135	39136
395014	**SB**	E	*SE*	AD	39141	39142	39143	39144	39145	39146
395015	**SB**	E	*SE*	AD	39151	39152	39153	39154	39155	39156
395016	**SB**	E	*SE*	AD	39161	39162	39163	39164	39165	39166
395017	**SB**	E	*SE*	AD	39171	39172	39173	39174	39175	39176
395018	**SB**	E	*SE*	AD	39181	39182	39183	39184	39185	39186
395019	**SB**	E	*SE*	AD	39191	39192	39193	39194	39195	39196
395020	**SB**	E	*SE*	AD	39201	39202	39203	39204	39205	39206
395021	**SB**	E	*SE*	AD	39211	39212	39213	39214	39215	39216
395022	**SB**	E	*SE*	AD	39221	39222	39223	39224	39225	39226
395023	**SB**	E	*SE*	AD	39231	39232	39233	39234	39235	39236
395024	**SB**	E	*SE*	AD	39241	39242	39243	39244	39245	39246
395025	**SB**	E	*SE*	AD	39251	39252	39253	39254	39255	39256
395026	**SB**	E	*SE*	AD	39261	39262	39263	39264	39265	39266
395027	**SB**	E	*SE*	AD	39271	39272	39273	39274	39275	39276
395028	**SB**	E	*SE*	AD	39281	39282	39283	39284	39285	39286
395029	**SB**	E	*SE*	AD	39291	39292	39293	39294	39295	39296

Names (carried on end cars):

395001	Dame Kelly Holmes	395007	Steve Backley
395002	Sebastian Coe	395008	Ben Ainslie
395003	Sir Steve Redgrave	395009	Rebecca Adlington
395004	Sir Chris Hoy	395016	Jamie Staff
395005	Dame Tanni Grey-Thompson	395 026	Marc Woods
395006	Daley Thompson		

2. 750 V DC THIRD RAIL EMUs

These classes use the third rail system at 750 V DC (unless stated). Outer couplers are buckeyes on units built before 1982 with bar couplers within the units. Newer units generally have Dellner outer couplers.

CLASS 442   WESSEX EXPRESS   BREL DERBY

Stock built for Waterloo–Bournemouth–Weymouth services. Previously used by South West Trains, all units now used by Southern, principally on Victoria–Gatwick Airport–Brighton services.

Formation: DTSO(A)–TSO–MBC–TSO(W)–DTSO(B).
Construction: Steel.
Traction Motors: Four EE546 of 300 kW recovered from Class 432s.
Wheel Arrangement: 2-2 + 2-2 + Bo-Bo + 2-2 + 2-2.
Braking: Disc.   **Dimensions:** 23.15/23.00 x 2.74 m.
Bogies: Two BREL P7 motor bogies (MBSO). T4 bogies (trailer cars).
Couplers: Buckeye.   **Control System:** 1986-type.
Gangways: Throughout.   **Maximum Speed:** 100 mph.
Doors: Sliding plug.   **Heating & Ventilation:** Air conditioning.
Seating Layout: 1: 2+1 facing, 2: 2+2 mainly unidirectional.
Multiple Working: Within class and Class 33/1 & 73 locos in an emergency.

DTSO(A). Lot No. 31030 Derby 1988–89. –/74. 38.5 t.
TSO. Lot No. 31032 Derby 1988–89. –/76 2T. 37.5 t.
MBC. Lot No. 31034 Derby 1988–89. 24/28. 55.0 t.
TSO(W). Lot No. 31033 Derby 1988–89. –/66(4) 1TD 1T 2W. 37.8 t.
DTSO(B). Lot No. 31031 Derby 1988–89. –/74. 37.3 t.

442401	GV	A	SN	SL	77382	71818	62937	71842	77406
442402	GV	A	SN	SL	77383	71819	62938	71843	77407
442403	GV	A	SN	SL	77384	71820	62941	71844	77408
442404	GV	A	SN	SL	77385	71821	62939	71845	77409
442405	GV	A	SN	SL	77386	71822	62944	71846	77410
442406	GV	A	SN	SL	77389	71823	62942	71847	77411
442407	GV	A	SN	SL	77388	71824	62943	71848	77412
442408	GV	A	SN	SL	77387	71825	62945	71849	77413
442409	GV	A	SN	SL	77390	71826	62946	71850	77414
442410	GV	A	SN	SL	77391	71827	62948	71851	77415
442411	GV	A	SN	SL	77392	71828	62940	71858	77422
442412	GV	A	SN	SL	77393	71829	62947	71853	77417
442413	GV	A	SN	SL	77394	71830	62949	71854	77418
442414	GV	A	SN	SL	77395	71831	62950	71855	77419
442415	GV	A	SN	SL	77396	71832	62951	71856	77420
442416	GV	A	SN	SL	77397	71833	62952	71857	77421
442417	GV	A	SN	SL	77398	71834	62953	71852	77416
442418	GV	A	SN	SL	77399	71835	62954	71859	77423
442419	GV	A	SN	SL	77400	71836	62955	71860	77424
442420	GV	A	SN	SL	77401	71837	62956	71861	77425
442421	GV	A	SN	SL	77402	71838	62957	71862	77426

442 422	**GV**	A	*SN*	SL	77403 71839 62958 71863 77427
442 423	**GV**	A	*SN*	SL	77404 71840 62959 71864 77428
442 424	**GV**	A	*SN*	SL	77405 71841 62960 71865 77429

CLASS 444   DESIRO UK   SIEMENS

Express units.

Formation: DMCO–TSO–TSO–TSORMB–DMSO.
Construction: Aluminium.
Traction Motors: 4 Siemens 1TB2016-0GB02 asynchronous of 250 kW.
Wheel Arrangement: Bo-Bo + 2-2 + 2-2 + 2-2 + Bo-Bo.
Braking: Disc, rheostatic & regenerative. **Dimensions:** 23.57 x 2.80 m.
Bogies: SGP SF5000.   **Couplers:** Dellner 12.
Gangways: Throughout.   **Control System:** IGBT Inverter.
Doors: Single-leaf sliding plug.   **Maximum Speed:** 100 mph.
Heating & Ventilation: Air conditioning.
Seating Layout: 1: 2+1 facing/unidirectional, 2: 2+2 facing/unidirectional.
Multiple Working: Within class and with Class 450.

DMSO. Siemens Vienna/Krefeld 2003–04. –/76. 51.3t.
TSO 67101–145. Siemens Vienna/Krefeld 2003–04. –/76 1T. 40.3t.
TSO 67151–195. Siemens Vienna/Krefeld 2003–04. –/76 1T. 36.8t.
TSORMB. Siemens Vienna/Krefeld 2003–04. –/47 1T 1TD 2W. 42.1t.
DMCO. Siemens Vienna/Krefeld 2003–04. 35/24. 51.3t.

444 001	**ST**	A	*SW*	NT	63801 67101 67151 67201 63851
444 002	**ST**	A	*SW*	NT	63802 67102 67152 67202 63852
444 003	**ST**	A	*SW*	NT	63803 67103 67153 67203 63853
444 004	**ST**	A	*SW*	NT	63804 67104 67154 67204 63854
444 005	**ST**	A	*SW*	NT	63805 67105 67155 67205 63855
444 006	**ST**	A	*SW*	NT	63806 67106 67156 67206 63856
444 007	**ST**	A	*SW*	NT	63807 67107 67157 67207 63857
444 008	**ST**	A	*SW*	NT	63808 67108 67158 67208 63858
444 009	**ST**	A	*SW*	NT	63809 67109 67159 67209 63859
444 010	**ST**	A	*SW*	NT	63810 67110 67160 67210 63860
444 011	**ST**	A	*SW*	NT	63811 67111 67161 67211 63861
444 012	**ST**	A	*SW*	NT	63812 67112 67162 67212 63862
444 013	**ST**	A	*SW*	NT	63813 67113 67163 67213 63863
444 014	**ST**	A	*SW*	NT	63814 67114 67164 67214 63864
444 015	**ST**	A	*SW*	NT	63815 67115 67165 67215 63865
444 016	**ST**	A	*SW*	NT	63816 67116 67166 67216 63866
444 017	**ST**	A	*SW*	NT	63817 67117 67167 67217 63867
444 018	**ST**	A	*SW*	NT	63818 67118 67168 67218 63868
444 019	**ST**	A	*SW*	NT	63819 67119 67169 67219 63869
444 020	**ST**	A	*SW*	NT	63820 67120 67170 67220 63870
444 021	**ST**	A	*SW*	NT	63821 67121 67171 67221 63871
444 022	**ST**	A	*SW*	NT	63822 67122 67172 67222 63872
444 023	**ST**	A	*SW*	NT	63823 67123 67173 67223 63873
444 024	**ST**	A	*SW*	NT	63824 67124 67174 67224 63874
444 025	**ST**	A	*SW*	NT	63825 67125 67175 67225 63875
444 026	**ST**	A	*SW*	NT	63826 67126 67176 67226 63876

444027	**ST**	A	*SW*	NT	63827	67127	67177	67227	63877
444028	**ST**	A	*SW*	NT	63828	67128	67178	67228	63878
444029	**ST**	A	*SW*	NT	63829	67129	67179	67229	63879
444030	**ST**	A	*SW*	NT	63830	67130	67180	67230	63880
444031	**ST**	A	*SW*	NT	63831	67131	67181	67231	63881
444032	**ST**	A	*SW*	NT	63832	67132	67182	67232	63882
444033	**ST**	A	*SW*	NT	63833	67133	67183	67233	63883
444034	**ST**	A	*SW*	NT	63834	67134	67184	67234	63884
444035	**ST**	A	*SW*	NT	63835	67135	67185	67235	63885
444036	**ST**	A	*SW*	NT	63836	67136	67186	67236	63886
444037	**ST**	A	*SW*	NT	63837	67137	67187	67237	63887
444038	**ST**	A	*SW*	NT	63838	67138	67188	67238	63888
444039	**ST**	A	*SW*	NT	63839	67139	67189	67239	63889
444040	**ST**	A	*SW*	NT	63840	67140	67190	67240	63890
444041	**ST**	A	*SW*	NT	63841	67141	67191	67241	63891
444042	**ST**	A	*SW*	NT	63842	67142	67192	67242	63892
444043	**ST**	A	*SW*	NT	63843	67143	67193	67243	63893
444044	**ST**	A	*SW*	NT	63844	67144	67194	67244	63894
444045	**ST**	A	*SW*	NT	63845	67145	67195	67245	63895

Names (carried on TSORMB):

444001	NAOMI HOUSE	444018	THE FAB 444
444012	DESTINATION WEYMOUTH		

CLASS 450   DESIRO UK   SIEMENS

Outer suburban units.

Formation: DMSO–TCO–TSO–DMSO (DMSO–TSO–TCO–DMSO 450 111–127).
Construction: Aluminium.
Traction Motors: 4 Siemens 1TB2016-0GB02 asynchronous of 250 kW.
Wheel Arrangement: Bo-Bo + 2-2 + 2-2 + Bo-Bo.
Braking: Disc, rheostatic & regenerative. **Dimensions:** 20.34 x 2.79 m.
Bogies: SGP SF5000. **Couplers:** Dellner 12.
Gangways: Throughout. **Control System:** IGBT Inverter.
Doors: Sliding plug. **Maximum Speed:** 100 mph.
Heating & Ventilation: Air conditioning.
Seating Layout: 1: 2+2 facing/unidirectional, 2: 3+2 facing/unidirectional.
Multiple Working: Within class and with Class 444.

Class 450/0. Standard units.

DMSO(A). Siemens Krefeld/Vienna 2002–06. –/70. 48.0 t.
TCO. Siemens Krefeld/Vienna 2002–06. 24/32(4) 1T. 35.8 t.
TSO. Siemens Krefeld/Vienna 2002–06. –/61(9) 1TD 2W. 39.8 t.
DMSO(B). Siemens Krefeld/Vienna 2002–06. –/70. 48.6 t.

450001	**SD**	A	*SW*	NT	63201	64201	68101	63601
450002	**SD**	A	*SW*	NT	63202	64202	68102	63602
450003	**SD**	A	*SW*	NT	63203	64203	68103	63603
450004	**SD**	A	*SW*	NT	63204	64204	68104	63604
450005	**SD**	A	*SW*	NT	63205	64205	68105	63605
450006	**SD**	A	*SW*	NT	63206	64206	68106	63606

450 007	**SD**	A	*SW*	NT	63207	64207	68107	63607
450 008	**SD**	A	*SW*	NT	63208	64208	68108	63608
450 009	**SD**	A	*SW*	NT	63209	64209	68109	63609
450 010	**SD**	A	*SW*	NT	63210	64210	68110	63610
450 011	**SD**	A	*SW*	NT	63211	64211	68111	63611
450 012	**SD**	A	*SW*	NT	63212	64212	68112	63612
450 013	**SD**	A	*SW*	NT	63213	64213	68113	63613
450 014	**SD**	A	*SW*	NT	63214	64214	68114	63614
450 015	**SD**	A	*SW*	NT	63215	64215	68115	63615
450 016	**SD**	A	*SW*	NT	63216	64216	68116	63616
450 017	**SD**	A	*SW*	NT	63217	64217	68117	63617
450 018	**SD**	A	*SW*	NT	63218	64218	68118	63618
450 019	**SD**	A	*SW*	NT	63219	64219	68119	63619
450 020	**SD**	A	*SW*	NT	63220	64220	68120	63620
450 021	**SD**	A	*SW*	NT	63221	64221	68121	63621
450 022	**SD**	A	*SW*	NT	63222	64222	68122	63622
450 023	**SD**	A	*SW*	NT	63223	64223	68123	63623
450 024	**SD**	A	*SW*	NT	63224	64224	68124	63624
450 025	**SD**	A	*SW*	NT	63225	64225	68125	63625
450 026	**SD**	A	*SW*	NT	63226	64226	68126	63626
450 027	**SD**	A	*SW*	NT	63227	64227	68127	63627
450 028	**SD**	A	*SW*	NT	63228	64228	68128	63628
450 029	**SD**	A	*SW*	NT	63229	64229	68129	63629
450 030	**SD**	A	*SW*	NT	63230	64230	68130	63630
450 031	**SD**	A	*SW*	NT	63231	64231	68131	63631
450 032	**SD**	A	*SW*	NT	63232	64232	68132	63632
450 033	**SD**	A	*SW*	NT	63233	64233	68133	63633
450 034	**SD**	A	*SW*	NT	63234	64234	68134	63634
450 035	**SD**	A	*SW*	NT	63235	64235	68135	63635
450 036	**SD**	A	*SW*	NT	63236	64236	68136	63636
450 037	**SD**	A	*SW*	NT	63237	64237	68137	63637
450 038	**SD**	A	*SW*	NT	63238	64238	68138	63638
450 039	**SD**	A	*SW*	NT	63239	64239	68139	63639
450 040	**SD**	A	*SW*	NT	63240	64240	68140	63640
450 041	**SD**	A	*SW*	NT	63241	64241	68141	63641
450 042	**SD**	A	*SW*	NT	63242	64242	68142	63642
450 071	**SD**	A	*SW*	NT	63271	64271	68171	63671
450 072	**SD**	A	*SW*	NT	63272	64272	68172	63672
450 073	**SD**	A	*SW*	NT	63273	64273	68173	63673
450 074	**SD**	A	*SW*	NT	63274	64274	68174	63674
450 075	**SD**	A	*SW*	NT	63275	64275	68175	63675
450 076	**SD**	A	*SW*	NT	63276	64276	68176	63676
450 077	**SD**	A	*SW*	NT	63277	64277	68177	63677
450 078	**SD**	A	*SW*	NT	63278	64278	68178	63678
450 079	**SD**	A	*SW*	NT	63279	64279	68179	63679
450 080	**SD**	A	*SW*	NT	63280	64280	68180	63680
450 081	**SD**	A	*SW*	NT	63281	64281	68181	63681
450 082	**SD**	A	*SW*	NT	63282	64282	68182	63682
450 083	**SD**	A	*SW*	NT	63283	64283	68183	63683
450 084	**SD**	A	*SW*	NT	63284	64284	68184	63684
450 085	**SD**	A	*SW*	NT	63285	64285	68185	63685

450086	**SD**	A	*SW*	NT	63286	64286	68186	63686
450087	**SD**	A	*SW*	NT	63287	64287	68187	63687
450088	**SD**	A	*SW*	NT	63288	64288	68188	63688
450089	**SD**	A	*SW*	NT	63289	64289	68189	63689
450090	**SD**	A	*SW*	NT	63290	64290	68190	63690
450091	**SD**	A	*SW*	NT	63291	64291	68191	63691
450092	**SD**	A	*SW*	NT	63292	64292	68192	63692
450093	**SD**	A	*SW*	NT	63293	64293	68193	63693
450094	**SD**	A	*SW*	NT	63294	64294	68194	63694
450095	**SD**	A	*SW*	NT	63295	64295	68195	63695
450096	**SD**	A	*SW*	NT	63296	64296	68196	63696
450097	**SD**	A	*SW*	NT	63297	64297	68197	63697
450098	**SD**	A	*SW*	NT	63298	64298	68198	63698
450099	**SD**	A	*SW*	NT	63299	64299	68199	63699
450100	**SD**	A	*SW*	NT	63300	64300	68200	63700
450101	**SD**	A	*SW*	NT	63701	66851	66801	63751
450102	**SD**	A	*SW*	NT	63702	66852	66802	63752
450103	**SD**	A	*SW*	NT	63703	66853	66803	63753
450104	**SD**	A	*SW*	NT	63704	66854	66804	63754
450105	**SD**	A	*SW*	NT	63705	66855	66805	63755
450106	**SD**	A	*SW*	NT	63706	66856	66806	63756
450107	**SD**	A	*SW*	NT	63707	66857	66807	63757
450108	**SD**	A	*SW*	NT	63708	66858	66808	63758
450109	**SD**	A	*SW*	NT	63709	66859	66809	63759
450110	**SD**	A	*SW*	NT	63710	66860	66810	63760
450111	**SD**	A	*SW*	NT	63901	66921	66901	63921
450112	**SD**	A	*SW*	NT	63902	66922	66902	63922
450113	**SD**	A	*SW*	NT	63903	66923	66903	63923
450114	**SD**	A	*SW*	NT	63904	66924	66904	63924
450115	**SD**	A	*SW*	NT	63905	66925	66905	63925
450116	**SD**	A	*SW*	NT	63906	66926	66906	63926
450117	**SD**	A	*SW*	NT	63907	66927	66907	63927
450118	**SD**	A	*SW*	NT	63908	66928	66908	63928
450119	**SD**	A	*SW*	NT	63909	66929	66909	63929
450120	**SD**	A	*SW*	NT	63910	66930	66910	63930
450121	**SD**	A	*SW*	NT	63911	66931	66911	63931
450122	**SD**	A	*SW*	NT	63912	66932	66912	63932
450123	**SD**	A	*SW*	NT	63913	66933	66913	63933
450124	**SD**	A	*SW*	NT	63914	66934	66914	63934
450125	**SD**	A	*SW*	NT	63915	66935	66915	63935
450126	**SD**	A	*SW*	NT	63916	66936	66916	63936
450127	**SD**	A	*SW*	NT	63917	66937	66917	63937

Names (carried on DMSO(B)):

| 450015 DESIRO | 450114 FAIRBRIDGE investing in the future |
| 450042 TRELOAR COLLEGE | |

Class 450/5. "High density" units. 28 units converted at Bournemouth for Waterloo–Windsor/Weybridge/Hounslow services. First Class removed and modified seating layout with more standing room. Details as Class 450/0 except:

Formation: DMSO–TSO–TSO–DMSO.

DMSO(A). Siemens Krefeld/Vienna 2002–04. –/64. 48.0 t.
TSO(A). Siemens Krefeld/Vienna 2002–04. –/56(4) 1T. 35.5 t.
TSO(B). Siemens Krefeld/Vienna 2002–04. –/56(9) 1TD 2W. 39.8 t.
DMSO(B). Siemens Krefeld/Vienna 2002–04. –/64. 48.6 t.

450543	(450043)	**SD**	A	*SW*	NT	63243	64243	68143 63643
450544	(450044)	**SD**	A	*SW*	NT	63244	64244	68144 63644
450545	(450045)	**SD**	A	*SW*	NT	63245	64245	68145 63645
450546	(450046)	**SD**	A	*SW*	NT	63246	64246	68146 63646
450547	(450047)	**SD**	A	*SW*	NT	63247	64247	68147 63647
450548	(450048)	**SD**	A	*SW*	NT	63248	64248	68148 63648
450549	(450049)	**SD**	A	*SW*	NT	63249	64249	68149 63649
450550	(450050)	**SD**	A	*SW*	NT	63250	64250	68150 63650
450551	(450051)	**SD**	A	*SW*	NT	63251	64251	68151 63651
450552	(450052)	**SD**	A	*SW*	NT	63252	64252	68152 63652
450553	(450053)	**SD**	A	*SW*	NT	63253	64253	68153 63653
450554	(450054)	**SD**	A	*SW*	NT	63254	64254	68154 63654
450555	(450055)	**SD**	A	*SW*	NT	63255	64255	68155 63655
450556	(450056)	**SD**	A	*SW*	NT	63256	64256	68156 63656
450557	(450057)	**SD**	A	*SW*	NT	63257	64257	68157 63657
450558	(450058)	**SD**	A	*SW*	NT	63258	64258	68158 63658
450559	(450059)	**SD**	A	*SW*	NT	63259	64259	68159 63659
450560	(450060)	**SD**	A	*SW*	NT	63260	64260	68160 63660
450561	(450061)	**SD**	A	*SW*	NT	63261	64261	68161 63661
450562	(450062)	**SD**	A	*SW*	NT	63262	64262	68162 63662
450563	(450063)	**SD**	A	*SW*	NT	63263	64263	68163 63663
450564	(450064)	**SD**	A	*SW*	NT	63264	64264	68164 63664
450565	(450065)	**SD**	A	*SW*	NT	63265	64265	68165 63665
450566	(450066)	**SD**	A	*SW*	NT	63266	64266	68166 63666
450567	(450067)	**SD**	A	*SW*	NT	63267	64267	68167 63667
450568	(450068)	**SD**	A	*SW*	NT	63268	64268	68168 63668
450569	(450069)	**SD**	A	*SW*	NT	63269	64269	68169 63669
450570	(450070)	**SD**	A	*SW*	NT	63270	64270	68170 63670

CLASS 455   BREL YORK

Inner suburban units.

Formation: DTSO–MSO–TSO–DTSO.
Construction: Steel. Class 455/7 TSO have a steel underframe and an aluminium alloy body & roof.
Traction Motors: Four GEC507-20J of 185 kW, some recovered from Class 405s.
Wheel Arrangement: 2-2 + Bo-Bo + 2-2 + 2-2.
Braking: Disc.   **Dimensions:** 19.92/19.83 x 2.82 m.
Bogies: P7 (motor) and T3 (455/8 & 455/9) BX1 (455/7) trailer.
Gangways: Within unit + end doors (sealed on Southern units).
Couplers: Tightlock.   **Control System:** 1982-type, camshaft.
Doors: Sliding.   **Maximum Speed:** 75 mph.
Heating & Ventilation: Various.
Seating Layout: All units refurbished. SWT units: 2+2 high-back unidirectional/ facing seating. Southern units: 3+2 high back mainly facing seating.

Multiple Working: Within class and with Class 456.

Class 455/7. South West Trains units. Second series with TSOs originally in Class 508s. Pressure heating & ventilation.

DTSO. Lot No. 30976 1984–85. –/50(4) 1W. 30.8t.
MSO. Lot No. 30975 1984–85. –/68. 45.7t.
TSO. Lot No. 30944 1979–80. –/68. 26.1t.

5701	**SS**	P	*SW*	WD	77727	62783	71545	77728
5702	**SS**	P	*SW*	WD	77729	62784	71547	77730
5703	**SS**	P	*SW*	WD	77731	62785	71540	77732
5704	**SS**	P	*SW*	WD	77733	62786	71548	77734
5705	**SS**	P	*SW*	WD	77735	62787	71565	77736
5706	**SS**	P	*SW*	WD	77737	62788	71534	77738
5707	**SS**	P	*SW*	WD	77739	62789	71536	77740
5708	**SS**	P	*SW*	WD	77741	62790	71560	77742
5709	**SS**	P	*SW*	WD	77743	62791	71532	77744
5710	**SS**	P	*SW*	WD	77745	62792	71566	77746
5711	**SS**	P	*SW*	WD	77747	62793	71542	77748
5712	**SS**	P	*SW*	WD	77749	62794	71546	77750
5713	**SS**	P	*SW*	WD	77751	62795	71567	77752
5714	**SS**	P	*SW*	WD	77753	62796	71539	77754
5715	**SS**	P	*SW*	WD	77755	62797	71535	77756
5716	**SS**	P	*SW*	WD	77757	62798	71564	77758
5717	**SS**	P	*SW*	WD	77759	62799	71528	77760
5718	**SS**	P	*SW*	WD	77761	62800	71557	77762
5719	**SS**	P	*SW*	WD	77763	62801	71558	77764
5720	**SS**	P	*SW*	WD	77765	62802	71568	77766
5721	**SS**	P	*SW*	WD	77767	62803	71553	77768
5722	**SS**	P	*SW*	WD	77769	62804	71533	77770
5723	**SS**	P	*SW*	WD	77771	62805	71526	77772
5724	**SS**	P	*SW*	WD	77773	62806	71561	77774
5725	**SS**	P	*SW*	WD	77775	62807	71541	77776
5726	**SS**	P	*SW*	WD	77777	62808	71556	77778
5727	**SS**	P	*SW*	WD	77779	62809	71562	77780
5728	**SS**	P	*SW*	WD	77781	62810	71527	77782
5729	**SS**	P	*SW*	WD	77783	62811	71550	77784
5730	**SS**	P	*SW*	WD	77785	62812	71551	77786
5731	**SS**	P	*SW*	WD	77787	62813	71555	77788
5732	**SS**	P	*SW*	WD	77789	62814	71552	77790
5733	**SS**	P	*SW*	WD	77791	62815	71549	77792
5734	**SS**	P	*SW*	WD	77793	62816	71531	77794
5735	**SS**	P	*SW*	WD	77795	62817	71563	77796
5736	**SS**	P	*SW*	WD	77797	62818	71554	77798
5737	**SS**	P	*SW*	WD	77799	62819	71544	77800
5738	**SS**	P	*SW*	WD	77801	62820	71529	77802
5739	**SS**	P	*SW*	WD	77803	62821	71537	77804
5740	**SS**	P	*SW*	WD	77805	62822	71530	77806
5741	**SS**	P	*SW*	WD	77807	62823	71559	77808
5742	**SS**	P	*SW*	WD	77809	62824	71543	77810
5750	**SS**	P	*SW*	WD	77811	62825	71538	77812

Class 455/8. Southern units. First series. Pressure heating & ventilation. Fitted with in-cab air conditioning systems meaning that the end door has been sealed.

DTSO. Lot No. 30972 York 1982–84. –/74. 33.6 t.
MSO. Lot No. 30973 York 1982–84. –/84. 37.9 t.
TSO. Lot No. 30974 York 1982–84. –/75(3) 2W. 34.0 t.

455801	**SN**	E	*SN*	SU	77627	62709	71657	77580
455802	**SN**	E	*SN*	SU	77581	62710	71664	77582
455803	**SN**	E	*SN*	SU	77583	62711	71639	77584
455804	**SN**	E	*SN*	SU	77585	62712	71640	77586
455805	**SN**	E	*SN*	SU	77587	62713	71641	77588
455806	**SN**	E	*SN*	SU	77589	62714	71642	77590
455807	**SN**	E	*SN*	SU	77591	62715	71643	77592
455808	**SN**	E	*SN*	SU	77637	62716	71644	77594
455809	**SN**	E	*SN*	SU	77623	62717	71648	77602
455810	**SN**	E	*SN*	SU	77597	62718	71646	77598
455811	**SN**	E	*SN*	SU	77599	62719	71647	77600
455812	**SN**	E	*SN*	SU	77595	62720	71645	77626
455813	**SN**	E	*SN*	SU	77603	62721	71649	77604
455814	**SN**	E	*SN*	SU	77605	62722	71650	77606
455815	**SN**	E	*SN*	SU	77607	62723	71651	77608
455816	**SN**	E	*SN*	SU	77609	62724	71652	77633
455817	**SN**	E	*SN*	SU	77611	62725	71653	77612
455818	**SN**	E	*SN*	SU	77613	62726	71654	77632
455819	**SN**	E	*SN*	SU	77615	62727	71637	77616
455820	**SN**	E	*SN*	SU	77617	62728	71656	77618
455821	**SN**	E	*SN*	SU	77619	62729	71655	77620
455822	**SN**	E	*SN*	SU	77621	62730	71658	77622
455823	**SN**	E	*SN*	SU	77601	62731	71659	77596
455824	**SN**	E	*SN*	SU	77593	62732	71660	77624
455825	**SN**	E	*SN*	SU	77579	62733	71661	77628
455826	**SN**	E	*SN*	SU	77630	62734	71662	77629
455827	**SN**	E	*SN*	SU	77610	62735	71663	77614
455828	**SN**	E	*SN*	SU	77631	62736	71638	77634
455829	**SN**	E	*SN*	SU	77635	62737	71665	77636
455830	**SN**	E	*SN*	SU	77625	62743	71666	77638
455831	**SN**	E	*SN*	SU	77639	62739	71667	77640
455832	**SN**	E	*SN*	SU	77641	62740	71668	77642
455833	**SN**	E	*SN*	SU	77643	62741	71669	77644
455834	**SN**	E	*SN*	SU	77645	62742	71670	77646
455835	**SN**	E	*SN*	SU	77647	62738	71671	77648
455836	**SN**	E	*SN*	SU	77649	62744	71672	77650
455837	**SN**	E	*SN*	SU	77651	62745	71673	77652
455838	**SN**	E	*SN*	SU	77653	62746	71674	77654
455839	**SN**	E	*SN*	SU	77655	62747	71675	77656
455840	**SN**	E	*SN*	SU	77657	62748	71676	77658
455841	**SN**	E	*SN*	SU	77659	62749	71677	77660
455842	**SN**	E	*SN*	SU	77661	62750	71678	77662
455843	**SN**	E	*SN*	SU	77663	62751	71679	77664
455844	**SN**	E	*SN*	SU	77665	62752	71680	77666
455845	**SN**	E	*SN*	SU	77667	62753	71681	77668

455846	**SN**	E	*SN*	SU	77669	62754	71682	77670

Class 455/8. South West Trains units. First series. Pressure heating & ventilation.

DTSO. Lot No. 30972 York 1982–84. –50(4) 1W. 29.5 t.
MSO. Lot No. 30973 York 1982–84. –/84 –/68. 45.6 t.
TSO. Lot No. 30974 York 1982–84. –/84 –/68. 27.1 t.

5847	**SS**	P	*SW*	WD	77671	62755	71683	77672
5848	**SS**	P	*SW*	WD	77673	62756	71684	77674
5849	**SS**	P	*SW*	WD	77675	62757	71685	77676
5850	**SS**	P	*SW*	WD	77677	62758	71686	77678
5851	**SS**	P	*SW*	WD	77679	62759	71687	77680
5852	**SS**	P	*SW*	WD	77681	62760	71688	77682
5853	**SS**	P	*SW*	WD	77683	62761	71689	77684
5854	**SS**	P	*SW*	WD	77685	62762	71690	77686
5855	**SS**	P	*SW*	WD	77687	62763	71691	77688
5856	**SS**	P	*SW*	WD	77689	62764	71692	77690
5857	**SS**	P	*SW*	WD	77691	62765	71693	77692
5858	**SS**	P	*SW*	WD	77693	62766	71694	77694
5859	**SS**	P	*SW*	WD	77695	62767	71695	77696
5860	**SS**	P	*SW*	WD	77697	62768	71696	77698
5861	**SS**	P	*SW*	WD	77699	62769	71697	77700
5862	**SS**	P	*SW*	WD	77701	62770	71698	77702
5863	**SS**	P	*SW*	WD	77703	62771	71699	77704
5864	**SS**	P	*SW*	WD	77705	62772	71700	77706
5865	**SS**	P	*SW*	WD	77707	62773	71701	77708
5866	**SS**	P	*SW*	WD	77709	62774	71702	77710
5867	**SS**	P	*SW*	WD	77711	62775	71703	77712
5868	**SS**	P	*SW*	WD	77713	62776	71704	77714
5869	**SS**	P	*SW*	WD	77715	62777	71705	77716
5870	**SS**	P	*SW*	WD	77717	62778	71706	77718
5871	**SS**	P	*SW*	WD	77719	62779	71707	77720
5872	**SS**	P	*SW*	WD	77721	62780	71708	77722
5873	**SS**	P	*SW*	WD	77723	62781	71709	77724
5874	**SS**	P	*SW*	WD	77725	62782	71710	77726

Class 455/9. South West Trains units. Third series. Convection heating.
Dimensions: 19.96/20.18 x 2.82 m.

DTSO. Lot No. 30991 York 1985. –/50(4) 1W. 30.7 t.
MSO. Lot No. 30992 York 1985. –/68. 46.3 t.
TSO. Lot No. 30993 York 1985. –/68. 28.3 t.
TSO†. Lot No. 30932 Derby 1981. –/68. 26.5 t.

Note: † Prototype vehicle 67400 converted from a Class 210 DEMU.

5901	**SS**	P	*SW*	WD	77813	62826	71714	77814
5902	**SS**	P	*SW*	WD	77815	62827	71715	77816
5903	**SS**	P	*SW*	WD	77817	62828	71716	77818
5904	**SS**	P	*SW*	WD	77819	62829	71717	77820
5905	**SS**	P	*SW*	WD	77821	62830	71725	77822
5906	**SS**	P	*SW*	WD	77823	62831	71719	77824
5907	**SS**	P	*SW*	WD	77825	62832	71720	77826

5908	**SS**	P	*SW*	WD	77827	62833	71721	77828
5909	**SS**	P	*SW*	WD	77829	62834	71722	77830
5910	**SS**	P	*SW*	WD	77831	62835	71723	77832
5911	**SS**	P	*SW*	WD	77833	62836	71724	77834
5912	† **SS**	P	*SW*	WD	77835	62837	67400	77836
5913	**SS**	P		ZN	77837	62838	71726	77838
5914	**SS**	P	*SW*	WD	77839	62839	71727	77840
5915	**SS**	P	*SW*	WD	77841	62840	71728	77842
5916	**SS**	P	*SW*	WD	77843	62841	71729	77844
5917	**SS**	P	*SW*	WD	77845	62842	71730	77846
5918	**SS**	P	*SW*	WD	77847	62843	71732	77848
5919	**SS**	P	*SW*	WD	77849	62844	71718	77850
5920	**SS**	P	*SW*	WD	77851	62845	71733	77852

CLASS 456   BREL YORK

Inner suburban units.

Formation: DMSO–DTSO.
Construction: Steel underframe, aluminium alloy body & roof.
Traction Motors: Two GEC507-20J of 185 kW, some recovered from Class 405s.
Wheel Arrangement: 2-Bo + 2-2.   **Dimensions:** 20.61 x 2.82 m.
Braking: Disc.   **Couplers:** Tightlock.
Bogies: P7 (motor) and T3 (trailer). **Control System:** GTO Chopper.
Gangways: Within unit.   **Maximum Speed:** 75 mph.
Doors: Sliding.   **Seating Layout:** 3+2 facing.
Heating & Ventilation: Convection heating.
Multiple Working: Within class and with Class 455.

DMSO. Lot No. 31073 1990–91. –/79. 41.1t.
DTSO. Lot No. 31074 1990–91. –/73. 31.4t.

Advertising livery: 456006 TfL/City of London (blue & green with various images).

456001	**SN**	P	*SN*	SU	64735	78250
456002	**SN**	P	*SN*	SU	64736	78251
456003	**SN**	P	*SN*	SU	64737	78252
456004	**SN**	P	*SN*	SU	64738	78253
456005	**SN**	P	*SN*	SU	64739	78254
456006	**AL**	P	*SN*	SU	64740	78255
456007	**SN**	P	*SN*	SU	64741	78256
456008	**SN**	P	*SN*	SU	64742	78257
456009	**SN**	P	*SN*	SU	64743	78258
456010	**SN**	P	*SN*	SU	64744	78259
456011	**SN**	P	*SN*	SU	64745	78260
456012	**SN**	P	*SN*	SU	64746	78261
456013	**SN**	P	*SN*	SU	64747	78262
456014	**SN**	P	*SN*	SU	64748	78263
456015	**SN**	P	*SN*	SU	64749	78264
456016	**SN**	P	*SN*	SU	64750	78265
456017	**SN**	P	*SN*	SU	64751	78266
456018	**SN**	P	*SN*	SU	64752	78267

456 019	**SN**	P	*SN*	SU	64753	78268
456 020	**SN**	P	*SN*	SU	64754	78269
456 021	**SN**	P	*SN*	SU	64755	78270
456 022	**SN**	P	*SN*	SU	64756	78271
456 023	**SN**	P	*SN*	SU	64757	78272
456 024	**SN**	P	*SN*	SU	64758	78273	Sir Cosmo Bonsor

CLASS 458   JUNIPER   ALSTOM BIRMINGHAM

Outer suburban units.

Formation: DMCO–TSO–MSO–DMCO.
Construction: Steel.
Traction Motors: Two Alstom ONIX 800 asynchronous of 270 kW.
Wheel Arrangement: 2-Bo + 2-2 + Bo-2 + Bo-2.
Braking: Disc & regenerative.   **Dimensions:** 21.16/19.94 x 2.80 m.
Bogies: ACR.   **Couplers:** Scharfenberg AAR.
Gangways: Throughout (not in use). **Control System:** IGBT Inverter.
Doors: Sliding plug.   **Maximum Speed:** 100 mph.
Heating & Ventilation: Air conditioning. **Multiple Working:** Within class.
Seating Layout: 1: 2+2 facing, 2: 3+2 facing/unidirectional.

DMCO(A). Alstom 1998–2000. 12/63. 46.4 t.
TSO. Alstom 1998–2000. –/54(6) 1TD 2W. 34.6 t.
MSO. Alstom 1998–2000. –/75 1T. 42.1 t.
DMCO(B). Alstom 1998–2000. 12/63. 46.4 t.

8001	**ST**	P	*SW*	WD	67601	74001	74101	67701
8002	**ST**	P	*SW*	WD	67602	74002	74102	67702
8003	**ST**	P	*SW*	WD	67603	74003	74103	67703
8004	**ST**	P	*SW*	WD	67604	74004	74104	67704
8005	**ST**	P	*SW*	WD	67605	74005	74105	67705
8006	**ST**	P	*SW*	WD	67606	74006	74106	67706
8007	**ST**	P	*SW*	WD	67607	74007	74107	67707
8008	**ST**	P	*SW*	WD	67608	74008	74108	67708
8009	**ST**	P	*SW*	WD	67609	74009	74109	67709
8010	**ST**	P	*SW*	WD	67610	74010	74110	67710
8011	**ST**	P	*SW*	WD	67611	74011	74111	67711
8012	**ST**	P	*SW*	WD	67612	74012	74112	67712
8013	**ST**	P	*SW*	WD	67613	74013	74113	67713
8014	**ST**	P	*SW*	WD	67614	74014	74114	67714
8015	**ST**	P	*SW*	WD	67615	74015	74115	67715
8016	**ST**	P	*SW*	WD	67616	74016	74116	67716
8017	**ST**	P	*SW*	WD	67617	74017	74117	67717
8018	**ST**	P	*SW*	WD	67618	74018	74118	67718
8019	**ST**	P	*SW*	WD	67619	74019	74119	67719
8020	**ST**	P	*SW*	WD	67620	74020	74120	67720
8021	**ST**	P	*SW*	WD	67621	74021	74121	67721
8022	**ST**	P	*SW*	WD	67622	74022	74122	67722
8023	**ST**	P	*SW*	WD	67623	74023	74123	67723
8024	**ST**	P	*SW*	WD	67624	74024	74124	67724
8025	**ST**	P	*SW*	WD	67625	74025	74125	67725

8026	**ST**	P	*SW*	WD	67626	74026	74126	67726
8027	**ST**	P	*SW*	WD	67627	74027	74127	67727
8028	**ST**	P	*SW*	WD	67628	74028	74128	67728
8029	**ST**	P	*SW*	WD	67629	74029	74129	67729
8030	**ST**	P	*SW*	WD	67630	74030	74130	67730

CLASS 460   JUNIPER   GEC-ALSTHOM

All units have now been stored and will be converted to 5-car Class 458s (all Class 458s will also be made 5-car sets).

Formation: DMLFO–TFO–TCO–MSO–MSO–TSO–MSO–DMSO.
Construction: Steel.
Traction Motors: Two Alstom ONIX 800 asynchronous of 270 kW.
Wheel Arrangement: 2-Bo + 2-2 + 2-2 +Bo-2 + 2-Bo + 2-2 + Bo-2 + Bo-2.
Braking: Disc & regenerative.   **Dimensions:** 21.01/19.94 x 2.80 m.
Bogies: ACR.
Couplers: Scharfenberg 330 at outer ends and between cars 4 and 5.
Gangways: Within unit.   **Control System:** IGBT Inverter.
Doors: Sliding plug.   **Maximum Speed:** 100 mph.
Heating & Ventilation: Air conditioning.
Seating Layout: 1: 2+1 facing, 2: 2+2 facing/unidirectional.
Multiple Working: Within class.

DMLFO. Alstom 1998–99. 10/– 42.7 t.
TFO. Alstom 1998–99. 25/– 1TD 1W. 34.5 t.
TCO. Alstom 1998–99. 8/38 1T. 35.6 t.
MSO(A). Alstom 1998–99. –/58. 42.8 t.
MSO(B). Alstom 1998–99. –/58. 42.5 t.
TSO. Alstom 1998–99. –/33 1TD 1T 1W. 35.2 t.
MSO(C). Alstom 1998–99. –/58. 40.5 t.
DMSO. Alstom 1998–99. –/54. 45.4 t.

Note: Vehicle 67904 of 460 004 is at LB and the other vehicles are at ZB.

460001	**GV**	P	SL	67901	74401	74411	74421
				74431	74441	74451	67911
460002	**GV**	P	SL	67902	74402	74412	74422
				74432	74442	74452	67912
460003	**GV**	P	BM	67903	74403	74413	74423
				74433	74443	74453	67913
460004	**GV**	P	ZB/LB	67904	74404	74414	74424
				74434	74444	74454	67914
460005	**GV**	P	SL	67905	74405	74415	74425
				74435	74445	74455	67915
460006	**GV**	P	BM	67906	74406	74416	74426
				74436	74446	74456	67916
460007	**GV**	P	SL	67907	74407	74417	74427
				74437	74447	74457	67917
460008	**GV**	P	BM	67908	74408	74418	74428
				74438	74448	74458	67918

CLASS 465 NETWORKER

Inner/outer suburban units.

Formation: DMSO–TSO–TSO–DMSO.
Construction: Welded aluminium alloy.
Traction Motors: Hitachi asynchronous of 280 kW (Classes 465/0 and 465/1) or GEC-Alsthom G352BY (Classes 465/2 and 465/9).
Wheel Arrangement: Bo-Bo + 2-2 + 2-2 + Bo-Bo.
Braking: Disc & rheostatic and regenerative (Classes 465/0 and 465/1 only).
Bogies: BREL P3/T3 (465/0 and 465/1), SRP BP62/BT52 (465/2 and 465/9).
Dimensions: 20.89/20.06 x 2.81 m.
Control System: IGBT Inverter (465/0 and 465/1) or 1992-type GTO Inverter.
Gangways: Within unit. **Couplers:** Tightlock.
Doors: Sliding plug. **Maximum Speed:** 75 mph.
Seating Layout: 3+2 facing/unidirectional.
Multiple Working: Within class and with Class 466.

64759–808. DMSO(A). Lot No. 31100 BREL York 1991–93. –/86. 39.2t.
64809–858. DMSO(B). Lot No. 31100 BREL York 1991–93. –/86. 39.2t.
65734–749. DMSO(A). Lot No. 31103 Metro-Cammell 1991–93. –/86. 39.2t.
65784–799. DMSO(B). Lot No. 31103 Metro-Cammell 1991–93. –/86. 39.2t.
65800–846. DMSO(A). Lot No. 31130 ABB York 1993–94. –/86. 39.2t.
65847–893. DMSO(B). Lot No. 31130 ABB York 1993–94. –/86. 39.2t.
72028–126 (even nos.) TSO. Lot No. 31102 BREL York 1991–93. –/90. 27.2t.
72029–127 (odd nos.) TSO. Lot No. 31101 BREL York 1991–93. –/86 1T. 28.0t.
72787–817 (odd nos.) TSO. Lot No. 31104 Metro-Cammell 1991–92. –/86 1T. 28.0t.
72788–818 (even nos.) TSO. Lot No. 31105 Metro-Cammell 1991–92. –/90. 27.2t.
72900–992 (even nos.) TSO. Lot No. 31102 ABB York 1993–94. –/90. 27.2t.
72901–993 (odd nos.) TSO. Lot No. 31101 ABB York 1993–94. –/86 1T. 28.0t.

Class 465/0. Built by BREL/ABB.

465001	**SE**	E	*SE*	SG	64759	72028	72029	64809
465002	**SE**	E	*SE*	SG	64760	72030	72031	64810
465003	**SE**	E	*SE*	SG	64761	72032	72033	64811
465004	**SE**	E	*SE*	SG	64762	72034	72035	64812
465005	**SE**	E	*SE*	SG	64763	72036	72037	64813
465006	**SE**	E	*SE*	SG	64764	72038	72039	64814
465007	**SE**	E	*SE*	SG	64765	72040	72041	64815
465008	**SE**	E	*SE*	SG	64766	72042	72043	64816
465009	**SE**	E	*SE*	SG	64767	72044	72045	64817
465010	**SE**	E	*SE*	SG	64768	72046	72047	64818
465011	**SE**	E	*SE*	SG	64769	72048	72049	64819
465012	**SE**	E	*SE*	SG	64770	72050	72051	64820
465013	**SE**	E	*SE*	SG	64771	72052	72053	64821
465014	**SE**	E	*SE*	SG	64772	72054	72055	64822
465015	**SE**	E	*SE*	SG	64773	72056	72057	64823
465016	**SE**	E	*SE*	SG	64774	72058	72059	64824
465017	**SE**	E	*SE*	SG	64775	72060	72061	64825
465018	**SE**	E	*SE*	SG	64776	72062	72063	64826
465019	**SE**	E	*SE*	SG	64777	72064	72065	64827

465020	**SE**	E	*SE*	SG	64778	72066	72067	64828
465021	**SE**	E	*SE*	SG	64779	72068	72069	64829
465022	**SE**	E	*SE*	SG	64780	72070	72071	64830
465023	**SE**	E	*SE*	SG	64781	72072	72073	64831
465024	**SE**	E	*SE*	SG	64782	72074	72075	64832
465025	**SE**	E	*SE*	SG	64783	72076	72077	64833
465026	**SE**	E	*SE*	SG	64784	72078	72079	64834
465027	**SE**	E	*SE*	SG	64785	72080	72081	64835
465028	**SE**	E	*SE*	SG	64786	72082	72083	64836
465029	**SE**	E	*SE*	SG	64787	72084	72085	64837
465030	**SE**	E	*SE*	SG	64788	72086	72087	64838
465031	**SE**	E	*SE*	SG	64789	72088	72089	64839
465032	**SE**	E	*SE*	SG	64790	72090	72091	64840
465033	**SE**	E	*SE*	SG	64791	72092	72093	64841
465034	**SE**	E	*SE*	SG	64792	72094	72095	64842
465035	**SE**	E	*SE*	SG	64793	72096	72097	64843
465036	**SE**	E	*SE*	SG	64794	72098	72099	64844
465037	**SE**	E	*SE*	SG	64795	72100	72101	64845
465038	**SE**	E	*SE*	SG	64796	72102	72103	64846
465039	**SE**	E	*SE*	SG	64797	72104	72105	64847
465040	**SE**	E	*SE*	SG	64798	72106	72107	64848
465041	**SE**	E	*SE*	SG	64799	72108	72109	64849
465042	**SE**	E	*SE*	SG	64800	72110	72111	64850
465043	**SE**	E	*SE*	SG	64801	72112	72113	64851
465044	**SE**	E	*SE*	SG	64802	72114	72115	64852
465045	**SE**	E	*SE*	SG	64803	72116	72117	64853
465046	**SE**	E	*SE*	SG	64804	72118	72119	64854
465047	**SE**	E	*SE*	SG	64805	72120	72121	64855
465048	**SE**	E	*SE*	SG	64806	72122	72123	64856
465049	**SE**	E	*SE*	SG	64807	72124	72125	64857
465050	**SE**	E	*SE*	SG	64808	72126	72127	64858

Class 465/1. Built by BREL/ABB. Similar to Class 465/0 but with detail differences.

465151	**SE**	E	*SE*	SG	65800	72900	72901	65847
465152	**SE**	E	*SE*	SG	65801	72902	72903	65848
465153	**SE**	E	*SE*	SG	65802	72904	72905	65849
465154	**SE**	E	*SE*	SG	65803	72906	72907	65850
465155	**SE**	E	*SE*	SG	65804	72908	72909	65851
465156	**SE**	E	*SE*	SG	65805	72910	72911	65852
465157	**SE**	E	*SE*	SG	65806	72912	72913	65853
465158	**SE**	E	*SE*	SG	65807	72914	72915	65854
465159	**SE**	E	*SE*	SG	65808	72916	72917	65855
465160	**SE**	E	*SE*	SG	65809	72918	72919	65856
465161	**SE**	E	*SE*	SG	65810	72920	72921	65857
465162	**SE**	E	*SE*	SG	65811	72922	72923	65858
465163	**SE**	E	*SE*	SG	65812	72924	72925	65859
465164	**SE**	E	*SE*	SG	65813	72926	72927	65860
465165	**SE**	E	*SE*	SG	65814	72928	72929	65861
465166	**SE**	E	*SE*	SG	65815	72930	72931	65862
465167	**SE**	E	*SE*	SG	65816	72932	72933	65863
465168	**SE**	E	*SE*	SG	65817	72934	72935	65864

465 169	**SE**	E	*SE*	SG	65818	72936	72937	65865
465 170	**SE**	E	*SE*	SG	65819	72938	72939	65866
465 171	**SE**	E	*SE*	SG	65820	72940	72941	65867
465 172	**SE**	E	*SE*	SG	65821	72942	72943	65868
465 173	**SE**	E	*SE*	SG	65822	72944	72945	65869
465 174	**SE**	E	*SE*	SG	65823	72946	72947	65870
465 175	**SE**	E	*SE*	SG	65824	72948	72949	65871
465 176	**SE**	E	*SE*	SG	65825	72950	72951	65872
465 177	**SE**	E	*SE*	SG	65826	72952	72953	65873
465 178	**SE**	E	*SE*	SG	65827	72954	72955	65874
465 179	**SE**	E	*SE*	SG	65828	72956	72957	65875
465 180	**SE**	E	*SE*	SG	65829	72958	72959	65876
465 181	**SE**	E	*SE*	SG	65830	72960	72961	65877
465 182	**SE**	E	*SE*	SG	65831	72962	72963	65878
465 183	**SE**	E	*SE*	SG	65832	72964	72965	65879
465 184	**SE**	E	*SE*	SG	65833	72966	72967	65880
465 185	**SE**	E	*SE*	SG	65834	72968	72969	65881
465 186	**SE**	E	*SE*	SG	65835	72970	72971	65882
465 187	**SE**	E	*SE*	SG	65836	72972	72973	65883
465 188	**SE**	E	*SE*	SG	65837	72974	72975	65884
465 189	**SE**	E	*SE*	SG	65838	72976	72977	65885
465 190	**SE**	E	*SE*	SG	65839	72978	72979	65886
465 191	**SE**	E	*SE*	SG	65840	72980	72981	65887
465 192	**SE**	E	*SE*	SG	65841	72982	72983	65888
465 193	**SE**	E	*SE*	SG	65842	72984	72985	65889
465 194	**SE**	E	*SE*	SG	65843	72986	72987	65890
465 195	**SE**	E	*SE*	SG	65844	72988	72989	65891
465 196	**SE**	E	*SE*	SG	65845	72990	72991	65892
465 197	**SE**	E	*SE*	SG	65846	72992	72993	65893

Class 465/2. Built by Metro-Cammell.
Dimensions: 20.80/20.15 x 2.81 m.

465 235	**SE**	A	*SE*	SG	65734	72787	72788	65784
465 236	**CN**	A	*SE*	SG	65735	72789	72790	65785
465 237	**SE**	A	*SE*	SG	65736	72791	72792	65786
465 238	**CN**	A	*SE*	SG	65737	72793	72794	65787
465 239	**CN**	A	*SE*	SG	65738	72795	72796	65788
465 240	**SE**	A	*SE*	SG	65739	72797	72798	65789
465 241	**SE**	A	*SE*	SG	65740	72799	72800	65790
465 242	**CN**	A	*SE*	SG	65741	72801	72802	65791
465 243	**SE**	A	*SE*	SG	65742	72803	72804	65792
465 244	**CN**	A	*SE*	SG	65743	72805	72806	65793
465 245	**CN**	A	*SE*	SG	65744	72807	72808	65794
465 246	**CN**	A	*SE*	SG	65745	72809	72810	65795
465 247	**CN**	A	*SE*	SG	65746	72811	72812	65796
465 248	**CN**	A	*SE*	SG	65747	72813	72814	65797
465 249	**CN**	A	*SE*	SG	65748	72815	72816	65798
465 250	**CN**	A	*SE*	SG	65749	72817	72818	65799

Class 465/9. Built by Metro-Cammell. Refurbished 2005 for longer distance services, with the addition of First Class seats. Details as Class 465/0 unless stated.

Formation: DMCO–TSO(A)–TSO(B)–DMCO.
Seating Layout: 1: 2+2 facing/unidirectional, 2: 3+2 facing/unidirectional.

65700–733. DMCO(A). Lot No. 31103 Metro-Cammell 1991–93. 12/68. 39.2t.
72719–785 (odd nos.) TSO(A). Lot No. 31104 Metro-Cammell 1991–92. –/76 1T 2W. 30.3t.
72720–786 (even nos.) TSO(B). Lot No. 31105 Metro-Cammell 1991–92. –/90. 29.5t.
65750–783. DMCO(B). Lot No. 31103 Metro-Cammell 1991–93. 12/68. 39.2t.

465901	(465201)	**CN**	A	*SE*	SG	65700	72719	72720	65750
465902	(465202)	**CN**	A	*SE*	SG	65701	72721	72722	65751
465903	(465203)	**CN**	A	*SE*	SG	65702	72723	72724	65752
465904	(465204)	**CN**	A	*SE*	SG	65703	72725	72726	65753
465905	(465205)	**CN**	A	*SE*	SG	65704	72727	72728	65754
465906	(465206)	**CN**	A	*SE*	SG	65705	72729	72730	65755
465907	(465207)	**SE**	A	*SE*	SG	65706	72731	72732	65756
465908	(465208)	**CN**	A	*SE*	SG	65707	72733	72734	65757
465909	(465209)	**CN**	A	*SE*	SG	65708	72735	72736	65758
465910	(465210)	**SE**	A	*SE*	SG	65709	72737	72738	65759
465911	(465211)	**CN**	A	*SE*	SG	65710	72739	72740	65760
465912	(465212)	**CN**	A	*SE*	SG	65711	72741	72742	65761
465913	(465213)	**SE**	A	*SE*	SG	65712	72743	72744	65762
465914	(465214)	**CN**	A	*SE*	SG	65713	72745	72746	65763
465915	(465215)	**CN**	A	*SE*	SG	65714	72747	72748	65764
465916	(465216)	**SE**	A	*SE*	SG	65715	72749	72750	65765
465917	(465217)	**CN**	A	*SE*	SG	65716	72751	72752	65766
465918	(465218)	**CN**	A	*SE*	SG	65717	72753	72754	65767
465919	(465219)	**CN**	A	*SE*	SG	65718	72755	72756	65768
465920	(465220)	**CN**	A	*SE*	SG	65719	72757	72758	65769
465921	(465221)	**SE**	A	*SE*	SG	65720	72759	72760	65770
465922	(465222)	**CN**	A	*SE*	SG	65721	72761	72762	65771
465923	(465223)	**SE**	A	*SE*	SG	65722	72763	72764	65772
465924	(465224)	**SE**	A	*SE*	SG	65723	72765	72766	65773
465925	(465225)	**CN**	A	*SE*	SG	65724	72767	72768	65774
465926	(465226)	**CN**	A	*SE*	SG	65725	72769	72770	65775
465927	(465227)	**CN**	A	*SE*	SG	65726	72771	72772	65776
465928	(465228)	**CN**	A	*SE*	SG	65727	72773	72774	65777
465929	(465229)	**CN**	A	*SE*	SG	65728	72775	72776	65778
465930	(465230)	**CN**	A	*SE*	SG	65729	72777	72778	65779
465931	(465231)	**CN**	A	*SE*	SG	65730	72779	72780	65780
465932	(465232)	**CN**	A	*SE*	SG	65731	72781	72782	65781
465933	(465233)	**SE**	A	*SE*	SG	65732	72783	72784	65782
465934	(465234)	**CN**	A	*SE*	SG	65733	72785	72786	65783

Name: 465903 Remembrance

CLASS 466   NETWORKER   GEC-ALSTHOM

Inner/outer suburban units.

Formation: DMSO–DTSO.
Construction: Welded aluminium alloy.
Traction Motors: Two GEC-Alsthom G352AY asynchronous of 280kW.
Wheel Arrangement: Bo-Bo + 2-2. **Couplers**: Tightlock.

Braking: Disc, rheostatic & regen.   **Control System:** 1992-type GTO Inverter.
Dimensions: 20.80 x 2.80 m.   **Maximum Speed:** 75 mph.
Bogies: BREL P3/T3.   **Doors:** Sliding plug.
Gangways: Within unit.   **Seating Layout:** 3+2 facing/unidirectional.
Multiple Working: Within class and with Class 465.

DMSO. Lot No. 31128 Birmingham 1993–94. –/86. 40.6t.
DTSO. Lot No. 31129 Birmingham 1993–94. –/82 1T. 31.4t.

466001	**CN**	A	*SE*	SG	64860	78312
466002	**CN**	A	*SE*	SG	64861	78313
466003	**SE**	A	*SE*	SG	64862	78314
466004	**SE**	A	*SE*	SG	64863	78315
466005	**SE**	A	*SE*	SG	64864	78316
466006	**SE**	A	*SE*	SG	64865	78317
466007	**SE**	A	*SE*	SG	64866	78318
466008	**SE**	A	*SE*	SG	64867	78319
466009	**CN**	A	*SE*	SG	64868	78320
466010	**SE**	A	*SE*	SG	64869	78321
466011	**SE**	A	*SE*	SG	64870	78322
466012	**CN**	A	*SE*	SG	64871	78323
466013	**SE**	A	*SE*	SG	64872	78324
466014	**SE**	A	*SE*	SG	64873	78325
466015	**CN**	A	*SE*	SG	64874	78326
466016	**CN**	A	*SE*	SG	64875	78327
466017	**CN**	A	*SE*	SG	64876	78328
466018	**SE**	A	*SE*	SG	64877	78329
466019	**SE**	A	*SE*	SG	64878	78330
466020	**CN**	A	*SE*	SG	64879	78331
466021	**SE**	A	*SE*	SG	64880	78332
466022	**SE**	A	*SE*	SG	64881	78333
466023	**SE**	A	*SE*	SG	64882	78334
466024	**SE**	A	*SE*	SG	64883	78335
466025	**SE**	A	*SE*	SG	64884	78336
466026	**SE**	A	*SE*	SG	64885	78337
466027	**SE**	A	*SE*	SG	64886	78338
466028	**SE**	A	*SE*	SG	64887	78339
466029	**SE**	A	*SE*	SG	64888	78340
466030	**SE**	A	*SE*	SG	64889	78341
466031	**SE**	A	*SE*	SG	64890	78342
466032	**SE**	A	*SE*	SG	64891	78343
466033	**CN**	A	*SE*	SG	64892	78344
466034	**SE**	A	*SE*	SG	64893	78345
466035	**SE**	A	*SE*	SG	64894	78346
466036	**CN**	A	*SE*	SG	64895	78347
466037	**SE**	A	*SE*	SG	64896	78348
466038	**SE**	A	*SE*	SG	64897	78349
466039	**SE**	A	*SE*	SG	64898	78350
466040	**SE**	A	*SE*	SG	64899	78351
466041	**CN**	A	*SE*	SG	64900	78352
466042	**SE**	A	*SE*	SG	64901	78353
466043	**CN**	A	*SE*	SG	64902	78354

CLASS 483   METRO-CAMMELL

Built 1938 onwards for LTE. Converted 1989–90 for the Isle of Wight Line.

Formation: DMSO–DMSO.
System: 660 V DC third rail.
Construction: Steel.
Traction Motors: Two Crompton Parkinson/GEC/BTH LT100 of 125 kW.
Braking: Tread.   **Dimensions:** 16.15 x 2.69 m.
Bogies: LT design.   **Couplers:** Wedglock.
Gangways: None. End doors.
Control System: Pneumatic Camshaft Motor (PCM).
Doors: Sliding.   **Maximum Speed:** 45 mph.
Seating Layout: Longitudinal or 2+2 facing/unidirectional.
Multiple Working: Within class.
Notes: The last three numbers of the unit number only are carried.

Former London Underground numbers are shown in parentheses.

DMSO (A). Lot No. 31071. –/40. 27.4 t.
DMSO (B). Lot No. 31072. –/42. 27.4 t.

483002	**LT**	SW		RY	122	(10221)	225	(11142)	RAPTOR
483004	**LT**	SW	*SW*	RY	124	(10205)	224	(11205)	
483006	**LT**	SW	*SW*	RY	126	(10297)	226	(11297)	
483007	**LT**	SW	*SW*	RY	127	(10291)	227	(11291)	
483008	**LT**	SW	*SW*	RY	128	(10255)	228	(11255)	
483009	**LT**	SW	*SW*	RY	129	(10289)	229	(11229)	

CLASS 507   BREL YORK

Formation: BDMSO–TSO–DMSO.
Construction: Steel underframe, aluminium alloy body and roof.
Traction Motors: Four GEC G310AZ of 82.125 kW.
Wheel Arrangement: Bo-Bo + 2-2 + Bo-Bo.
Braking: Disc & rheostatic.   **Dimensions:** 20.18 x 2.82 m.
Bogies: BX1.   **Couplers:** Tightlock.
Gangways: Within unit + end doors.   **Control System:** Camshaft.
Doors: Sliding.   **Maximum Speed:** 75 mph.
Seating Layout: All refurbished with 2+2 high-back facing seating.
Multiple Working: Within class and with Class 508.

BDMSO. Lot No. 30906 1978–80. –/56(3) 1W. 37.0 t.
TSO. Lot No. 30907 1978–80. –/74. 25.5 t.
DMSO. Lot No. 30908 1978–80. –/56(3) 1W. 35.5 t.

Advertising livery: 507 002 Liverpool Hope University (white).

507001	**ME**	A	*ME*	BD	64367	71342	64405
507002	**AL**	A	*ME*	BD	64368	71343	64406
507003	**ME**	A	*ME*	BD	64369	71344	64407
507004	**ME**	A	*ME*	BD	64388	71345	64408	Bob Paisley
507005	**ME**	A	*ME*	BD	64371	71346	64409
507006	**ME**	A	*ME*	BD	64372	71347	64410

507007	**ME**	A	*ME*	BD	64373	71348	64411	
507008	**ME**	A	*ME*	BD	64374	71349	64412	
507009	**ME**	A	*ME*	BD	64375	71350	64413	Dixie Dean
507010	**ME**	A	*ME*	BD	64376	71351	64414	
507011	**ME**	A	*ME*	BD	64377	71352	64415	
507012	**ME**	A	*ME*	BD	64378	71353	64416	
507013	**ME**	A	*ME*	BD	64379	71354	64417	
507014	**ME**	A	*ME*	BD	64380	71355	64418	
507015	**ME**	A	*ME*	BD	64381	71356	64419	
507016	**ME**	A	*ME*	BD	64382	71357	64420	
507017	**ME**	A	*ME*	BD	64383	71358	64421	
507018	**ME**	A	*ME*	BD	64384	71359	64422	
507019	**ME**	A	*ME*	BD	64385	71360	64423	
507020	**ME**	A	*ME*	BD	64386	71361	64424	John Peel
507021	**ME**	A	*ME*	BD	64387	71362	64425	Red Rum
507023	**ME**	A	*ME*	BD	64389	71364	64427	Operations Inspector Stuart Mason
507024	**ME**	A	*ME*	BD	64390	71365	64428	
507025	**ME**	A	*ME*	BD	64391	71366	64429	
507026	**ME**	A	*ME*	BD	64392	71367	64430	
507027	**ME**	A	*ME*	BD	64393	71368	64431	
507028	**ME**	A	*ME*	BD	64394	71369	64432	
507029	**ME**	A	*ME*	BD	64395	71370	64433	
507030	**ME**	A	*ME*	BD	64396	71371	64434	
507031	**ME**	A	*ME*	BD	64397	71372	64435	
507032	**ME**	A	*ME*	BD	64398	71373	64436	
507033	**ME**	A	*ME*	BD	64399	71374	64437	Councillor Jack Spriggs

CLASS 508   BREL YORK

Formation: DMSO–TSO–BDMSO.
Construction: Steel underframe, aluminium alloy body and roof.
Traction Motors: Four GEC G310AZ of 82.125 kW.
Wheel Arrangement: Bo-Bo + 2-2 + Bo-Bo.
Braking: Disc & rheostatic.   **Dimensions:** 20.18 x 2.82 m.
Bogies: BX1.   **Couplers:** Tightlock.
Gangways: Within unit + end doors. **Control System:** Camshaft.
Doors: Sliding.   **Maximum Speed:** 75 mph.
Seating Layout: All Merseyrail units refurbished with 2+2 high-back facing seating. 508/2 and 508/3 units have 3+2 low-back facing seating.
Multiple Working: Within class and with Class 507.

DMSO. Lot No. 30979 1979–80. –/56(3) 1W. 36.0 t.
TSO. Lot No. 30980 1979–80. –/74. 26.5 t.
BDMSO. Lot No. 30981 1979–80. –/56(3) 1W. 36.5 t.

Advertising livery: 508 111 Beatles Story (blue).

Class 508/1. Merseyrail units.

508103	**ME**	A	*ME*	BD	64651	71485	64694
508104	**ME**	A	*ME*	BD	64652	71486	64695
508108	**ME**	A	*ME*	BD	64656	71490	64699

508110	**ME**	A	*ME*	BD	64658	71492	64701
508111	**AL**	A	*ME*	BD	64659	71493	64702
508112	**ME**	A	*ME*	BD	64660	71494	64703
508114	**ME**	A	*ME*	BD	64662	71496	64705
508115	**ME**	A	*ME*	BD	64663	71497	64706
508117	**ME**	A	*ME*	BD	64665	71499	64708
508120	**ME**	A	*ME*	BD	64668	71502	64711
508122	**ME**	A	*ME*	BD	64670	71504	64713
508123	**ME**	A	*ME*	BD	64671	71505	64714
508124	**ME**	A	*ME*	BD	64672	71506	64715
508125	**ME**	A	*ME*	BD	64673	71507	64716
508126	**ME**	A	*ME*	BD	64674	71508	64717
508127	**ME**	A	*ME*	BD	64675	71509	64718
508128	**ME**	A	*ME*	BD	64676	71510	64719
508130	**ME**	A	*ME*	BD	64678	71512	64721
508131	**ME**	A	*ME*	BD	64679	71513	64722
508134	**ME**	A	*ME*	BD	64682	71516	64725
508136	**ME**	A	*ME*	BD	64684	71518	64727
508137	**ME**	A	*ME*	BD	64685	71519	64728
508138	**ME**	A	*ME*	BD	64686	71520	64729
508139	**ME**	A	*ME*	BD	64687	71521	64730
508140	**ME**	A	*ME*	BD	64688	71522	64731
508141	**ME**	A	*ME*	BD	64689	71523	64732
508143	**ME**	A	*ME*	BD	64691	71525	64734

Class 508/2. Units facelifted for the South Eastern lines by Wessex Traincare/Alstom, Eastleigh 1998–99.

DMSO. Lot No. 30979 1979–80. –/66. 36.0 t.
TSO. Lot No. 30980 1979–80. –/79 1W. 26.5 t.
BDMSO. Lot No. 30981 1979–80. –/74. 36.5 t.

508201	(508101)	**CX**	A	ZG	64649	71483	64692
508202	(508105)	**CX**	A	ZG	64653	71487	64696
508203	(508106)	**CN**	A	ZG	64654	71488	64697
508204	(508107)	**CX**	A	ZG	64655	71489	64698
508205	(508109)	**CN**	A	ZG	64657	71491	64700
508206	(508113)	**CX**	A	ZG	64661	71495	64704
508207	(508116)	**CN**	A	ZG	64664	71498	64707
508208	(508119)	**CN**	A	ZG	64667	71501	64710
508209	(508121)	**CX**	A	ZG	64669	71503	64712
508210	(508129)	**CN**	A	ZG	64677	71515	64720
508211	(508132)	**CN**	A	ZG	64680	71514	64723

Class 508/3. Units facelifted units for use on Euston–Watford Junction services by Alstom, Eastleigh 2002–03.

DMSO. Lot No. 30979 1979–80. –/68 1W. 36.0 t.
TSO. Lot No. 30980 1979–80. –/86. 26.5 t.
BDMSO. Lot No. 30981 1979–80. –/68 1W. 36.5 t.

508301	(508102)	**SL**	A	ZG	64650	71484	64693
508302	(508135)	**SL**	A	ZG	64683	71517	64726
508303	(508142)	**SL**	A	ZG	64690	71524	64733

3. EUROSTAR UNITS (CLASS 373)

Eurostar units were built for and are normally used on services between Britain and continental Europe via the Channel Tunnel. SNCF-owned units 3203/04, 3225/26 and 3227/28 have been removed from the Eurostar pool and only operate SNCF internal services between Paris and Lille. As they are not now permitted through the Channel Tunnel they are not listed here.

Each train consists of two 10-car units coupled, with a motor car at each driving end. All units are articulated with an extra motor bogie on the coach adjacent to the motor car.

All sets can be used between London St Pancras and Paris, Brussels and Disneyland Paris. Certain sets (shown *) are equipped for 1500 V DC operation and are used for the winter service to Bourg Saint Maurice and the summer service to Avignon.

Seven 8-car sets were built for Regional Eurostar services, but all except one power car (3308) and one half set are on long-term hire to SNCF for use on French internal services so are not listed here. The spare half set (from 3308/07) is stored at Temple Mills depot.

Formation: DM–MSO–4TSO–RB–2TFO–TBFO. Gangwayed within pair of units. Air conditioned.
Construction: Steel.
Supply Systems: 25 kV AC 50 Hz overhead or 3000 V DC overhead (* also equipped for 1500 V DC overhead operation).
Control System: GTO–GTO Inverter on UK 750 V DC and 25 kV AC, GTO Chopper on SNCB 3000 V DC.
Wheel Arrangement: Bo–Bo + Bo–2–2–2–2–2–2–2–2.
Length: 22.15 m (DM), 21.85 m (MS & TBF), 18.70 m (other cars).
Couplers: Schaku 10S at outer ends, Schaku 10L at inner end of each DM and outer ends of each sub set.
Maximum Speed: 186 mph (300 km/h)
Built: 1992–93 by GEC-Alsthom/Brush/ANF/De Dietrich/BN Construction/ ACEC.
Note: DM vehicles carry the set numbers indicated below.

Class 373/0. 10-car sets. Built for services starting from/terminating in London Waterloo (now St Pancras). Individual vehicles in each set are allocated numbers 373xxx0 + 373xxx1 + 373xxx2 + 373xxx3 + 373xxx4 + 373xxx5 + 373xxx6 + 373xxx7 + 373xxx8 + 373xxx9, where 3xxx denotes the set number.

373xxx0 series. DM. Lot No. 31118 1992–95. 68.5 t.
373xxx1 series. MSO. Lot No. 31119 1992–95. –/48 2T. 44.6 t.
373xxx2 series. TSO. Lot No. 31120 1992–95. –/56 1T. 28.1 t.
373xxx3 series. TSO. Lot No. 31121 1992–95. –/56 2T. 29.7 t.
373xxx4 series. TSO. Lot No. 31122 1992–95. –/56 1T. 28.3 t.
373xxx5 series. TSO. Lot No. 31123 1992–95. –/56 2T. 29.2 t.
373xxx6 series. RB. Lot No.31124 1992–95. 31.1 t.
373xxx7 series. TFO. Lot No. 31125 1992–95. 39/– 1T. 29.6 t.
373xxx8 series. TFO. Lot No. 31126 1992–95. 39/– 1T. 32.2 t.
373xxx9 series. TBFO. Lot No. 31127 1992–95. 25/– 1TD. 39.4 t.

No.						No.					
3001		**EU**	EU	*EU*	TI	3107		**EU**	SB	*EU*	FF
3002		**EU**	EU	*EU*	TI	3108		**EU**	SB	*EU*	FF
3003		**EU**	EU	*EU*	TI	3201	*	**EU**	SF	*EU*	LY
3004		**EU**	EU	*EU*	TI	3202	*	**EU**	SF	*EU*	LY
3005		**EU**	EU	*EU*	TI	3205		**EU**	SF	*EU*	LY
3006		**EU**	EU	*EU*	TI	3206		**EU**	SF	*EU*	LY
3007		**EU**	EU	*EU*	TI	3207	*	**EU**	SF	*EU*	LY
3008		**EU**	EU	*EU*	TI	3208	*	**EU**	SF	*EU*	LY
3009		**EU**	EU	*EU*	TI	3209	*	**EU**	SF	*EU*	LY
3010		**EU**	EU	*EU*	TI	3210	*	**EU**	SF	*EU*	LY
3011		**EU**	EU	*EU*	TI	3211		**EU**	SF	*EU*	LY
3012		**EU**	EU	*EU*	TI	3212		**EU**	SF	*EU*	LY
3013		**EU**	EU	*EU*	TI	3213	*	**EU**	SF	*EU*	LY
3014		**EU**	EU	*EU*	TI	3214	*	**EU**	SF	*EU*	LY
3015		**EU**	EU	*EU*	TI	3215	*	**EU**	SF	*EU*	LY
3016		**EU**	EU	*EU*	TI	3216	*	**EU**	SF	*EU*	LY
3017		**EU**	EU	*EU*	TI	3217		**EU**	SF	*EU*	LY
3018		**EU**	EU	*EU*	TI	3218		**EU**	SF	*EU*	LY
3019		**EU**	EU	*EU*	TI	3219		**EU**	SF	*EU*	LY
3020		**EU**	EU	*EU*	TI	3220		**EU**	SF	*EU*	LY
3021		**EU**	EU	*EU*	TI	3221		**EU**	SF	*EU*	LY
3022		**EU**	EU	*EU*	TI	3222		**EU**	SF	*EU*	LY
3101		**EU**	SB		TI	3223	*	**EU**	SF	*EU*	LY
3102		**EU**	SB		TI	3224	*	**EU**	SF	*EU*	LY
3103		**EU**	SB	*EU*	FF	3229	*	**EU**	SF	*EU*	LY
3104		**EU**	SB	*EU*	FF	3230	*	**EU**	SF	*EU*	LY
3105		**EU**	SB	*EU*	FF	3231		**EU**	SF	*EU*	LY
3106		**EU**	SB	*EU*	FF	3232		**EU**	SF	*EU*	LY

Spare Regional Eurostar DM:

3308	**EU**	EU		LB

Spare DM:

3999	**EU**	EU	*EU*	TI

Names:

3001/02	Tread Lightly/Voyage Vert	3013/14	LONDON 2012
3003/04	Tri-City-Athlon 2010	3207/08	MICHEL HOLLARD
3007/08	Waterloo Sunset	3209/10	THE DA VINCI CODE
3009/10	REMEMBERING FROMELLES		

▲ Southern-liveried 377 211, one of the dual voltage units, leads an 8-car formation on the 15.03 London Bridge–Horsham at South Croydon on 12/04/11.
Brian Denton

▼ London Overground-liveried 378 216 is seen near Brondesbury with the 08.36 Richmond–Stratford on 25/07/11.
Brian Denton

▲ In National Express livery, 379 022 and 379 027 arrive at Harlow Town with the 11.28 London Liverpool Street–Cambridge on 22/08/11.　**Robert Pritchard**

▼ Class 380s in the 380 101–108 batch work the North Berwick branch. On 03/08/12 380 108 passes East Fenton on the single-track line with the 11.27 North Berwick–Edinburgh.　**Robert Pritchard**

▲ One of the new 11-car Pendolinos delivered in 2011–12, 390 154, is seen just south of Carstairs working the 15.40 Glasgow Central–London Euston on 11/08/12. **Robin Ralston**

▼ Southeastern blue-liveried 395 008 and 395 005 pass Rainham on High Speed 1 with an empty stock working from Ashford to London St Pancras on 23/03/11.
Robert Pritchard

▲ The former Wessex Electrics are now used on Gatwick Express services. On 15/03/12 442 417 and 442 419 pass Stoats Nest Jn, Coulsdon, working the 16.00 London Victoria–Gatwick Airport. **Robert Pritchard**

▼ South West Trains white-liveried 444 007 leads the 09.20 Weymouth–London Waterloo at Christchurch on 11/04/12. **Andrew Mist**

▲ SWT blue-liveried 450 110 passes Battledown, west of Basingstoke, with the 12.09 Waterloo–Portsmouth Harbour via Eastleigh on 18/01/11.　**Brian Denton**

▼ Southern-liveried 455 814 arrives at Selhurst with the 13.03 London Victoria–Sutton on 15/03/12.　**Robert Pritchard**

▲ Southern-liveried 456 019 stands at London Bridge on 12/04/11.
Brian Denton

▼ Southeastern-liveried 465 050 passes Clapham High Street with the 09.55 London Victoria–Orpington on 07/12/11.
Robert Pritchard

▲ In Merseyrail livery, 508 124 arrives at Rock Ferry with the 12.45 Liverpool Central–Chester on 06/11/11.　**Robert Pritchard**

▼ Bombardier-built M5000 Flexity Swift tram 3007 pauses at Old Trafford working a Bury–Altrincham service on 17/04/12.　**Robert Pritchard**

▲ One of the six new Stadler trams delivered to Croydon in 2012, 2557 leaves Arena with a Therapia Lane–Elmers End service on 17/07/12. **Murdoch Currie**

▼ Sheffield Supertram 112 leaves the Sheffield Station stop heading for Halfway with a blue route service on 15/09/12. **Robert Pritchard**

4. INTERNAL USE EMUS

The following two vehicles are used for staff training at Virgin's training centre in Crewe. They are from Pendolino 390033 which was damaged in the Lambrigg accident of February 2007.

(390033)	**VT**	VI	Crewe	69133	69833

5. EMUS AWAITING DISPOSAL

The list below comprises vehicles awaiting disposal which are stored on the national railway network.

25 kV AC 50 Hz OVERHEAD UNITS:

(390033)	**VT**	VI	LM	69433	69533

Spare cars:

Cl. 309	**RR**	WC	CS	71758
Cl. 365	**N**	FC	ZN	65919

750 V DC THIRD RAIL UNITS:

3905	**CX**	BT	AF	76398	62266	70904	76397
3918	**CX**	BT	AF	76528	62321	70950	76527
930 010	**B/RK**	EM	DY	975600	(10988)	975601	(10843)

Spare car:

Former Class 210 DEMU vehicle being assessed for use as a replacement for one of the damaged vehicles of Class 455 5913.

Cl. 210	**N**	E	ZN	67301

6. UK LIGHT RAIL & METRO SYSTEMS

This section lists the rolling stock of the various light rail and metro systems in Great Britain. Passenger carrying vehicles only are covered (not works vehicles). This listing does not cover the London Underground network.

6.1. BLACKPOOL & FLEETWOOD TRAMWAY

Until the opening of Manchester Metrolink, the Blackpool tramway was the only urban/inter-urban tramway system left in Britain. The infrastructure is owned by Blackpool Corporation, and the tramway is operated by Blackpool Transport Services Ltd. The 11½ miles from Fleetwood to Starr Gate reopened in spring 2012 as a modern light rail system with a new fleet of 16 Bombardier Flexity 2 trams.

System: 600 V DC overhead.
Depot & Workshops: Starr Gate and Rigby Road (heritage fleet).
Standard livery: Flexity 2s and **F**: White, black & purple.

FLEXITY 2   5-SECTION TRAMS

16 new articulated Supertrams. These form the "A" fleet of trams used in daily service.

Built: 2011–12 by Bombardier Transportation, Bautzen, Germany.
Wheel arrangement: Bo-2-Bo.
Traction Motors: Four Bombardier 3-phase asynchronous of 120 kW.
Dimensions: 32.2 x 2.65 m.   **Seats:** 70 (4).
Doors: Sliding plug.   **Couplers:**
Weight: 40.9 t.   **Maximum Speed:** 43 mph.
Braking: Regenerative, disc and magnetic track.

001	005	009	013
002	006	010	014
003	007	011	015
004	008	012	016

"BALLOON" DOUBLE DECKERS   A1-1A

The 11 cars listed have partial exemption from the Rail Vehicle Accessibility Regulations and form the "B" fleet to supplement the Flexity 2 trams. All except 701 and 723 have been fitted with wider doors.

Built: 1934–35 by English Electric.
Traction Motors: Two EE305 of 40 kW.   **Seats:** 94 (*† 92, ‡ 90).
Notes:
* Rebuilt with a flat front end design and air-conditioned cabs. Known as "Millennium Class".
719 is named "DONNA'S DREAM HOUSE".

Advertising liveries:

707 – Coral Island – The Jewel on the Mile (black)
709 – Blackpool Sealife Centre (blue)
711 – Blackpool Zoo (various)
713 – Houndshill Shopping Centre (purple & white)
718 – Madame Tussaud's (purple & red)
719 – Pleasure Beach Resort (black & gold)
720 – Walls ice cream (red)
723 – Sands Venue nightclub (black)
724 – Lyndene Hotel (blue)

700	**F**		711 †	**AL**		720		**AL**
701 ‡(S)	**Yellow**		713	**AL**		723 †(S)		**AL**
707 *	**AL**		718 *	**AL**		724 *		**AL**
709 *	**AL**		719	**AL**				

HERITAGE FLEET of VINTAGE CARS

The following trams have full exemption from the RVAR and form the "C" fleet for Heritage use.

Blackpool & Fleetwood 40	Single deck "box car"	Built: 1914
Bolton 66	Bogie double-decker	Built: 1901
Blackpool 147 MICHAEL AIREY	Standard double-decker	Built: 1924
Blackpool 230 (604) GEORGE FORMBY OBE	Open boat car	Built: 1934
Blackpool 272+T2 (672+682)	Progress Twin Car	Rebuilt: 1960
Blackpool 600 THE DUCHESS OF CORNWALL	Open boat car	Built: 1934
Blackpool 602 (S)	Open boat car	Built: 1934
Blackpool 631	Brush car	Built: 1937
Blackpool 648	Centenary car	Built: 1987
Blackpool 660 (S)	Coronation Class single-decker	Built: 1953
Blackpool 675+685 (S)	Progress Twin Car	Rebuilt: 1958–60
Blackpool 706 PRINCESS ALICE	Balloon open-top double decker	Built: 1934
Blackpool 717 PHILLIP R THORPE	Balloon double decker	Built: 1934

Illuminated cars

Blackpool 733	Western Train loco & tender	Rebuilt: 1962
Blackpool 734	Western Train coach	Rebuilt: 1962
Blackpool 736	"Warship" HMS Blackpool	Rebuilt: 1965
Blackpool 737	Illuminated Trawler – "Fisherman's Friend"	Rebuilt: 2001

STORED VEHICLES

The following vehicles are stored or awaiting disposal.

Blackpool 642	Centenary car	Built: 1986
Blackpool 676+686	Progress Twin Car	Rebuilt: 1958–60
Blackpool 680	English Electric Railcoach	Rebuilt 1960

6.2. DOCKLANDS LIGHT RAILWAY

This system runs for a total of approximately 23 route miles from termini at Bank and Tower Gateway in central London to Lewisham, Stratford, Beckton and Woolwich Arsenal. A new line from Canning Town to Stratford International also opened in 2011. The first part of the network opened in 1987 from Tower Gateway to Island Gardens. Originally owned by London Transport, it is now part of the London Rail division of Transport for London and operated by Serco Docklands. Cars are normally "driven" automatically using the Alcatel "Seltrack" moving block signalling system.

Notes: Original P86 and P89 Class vehicles 01–21 were withdrawn from service in 1991 (01–11) and 1995 (12–21) and sold for use in Essen, Germany.

55 new cars from Bombardier in Germany entered traffic between 2008 and 2010. Along with platform extensions, these new vehicles have enabled 3-unit trains to operate on all routes.

System: 750 V DC third rail (bottom contact). High-floor.
Depots: Beckton (main depot) and Poplar.
Livery: Red with a curving blue stripe to represent the River Thames.
Advertising liveries:
54, 62, 82 – Lycamobile (white)
63, 74, 94 – General Electric (black)

CLASS B90				2-SECTION UNITS	

Built: 1991–92 by BN Construction, Bruges, Belgium. Chopper control.
Wheel Arrangement: B-2-B.   **Traction Motors:** Two Brush of 140 kW.
Seats: 52 (4).   **Weight:** 37 t.
Dimensions: 28.80 x 2.65 m.   **Braking:** Rheostatic.
Couplers: Scharfenberg.   **Maximum Speed:** 50 mph.
Doors: Sliding. End doors for staff use.

22	26	30	34	38	42
23	27	31	35	39	43
24	28	32	36	40	44
25	29	33	37	41	

CLASS B92				2-SECTION UNITS	

Built: 1992–95 by BN Construction, Bruges, Belgium. Chopper control.
Wheel Arrangement: B-2-B.   **Traction Motors:** Two Brush of 140 kW.
Seats: 52 (4).   **Weight:** 37 t.
Dimensions: 28.80 x 2.65 m.   **Braking:** Rheostatic.
Couplers: Scharfenberg.   **Maximum Speed:** 50 mph.
Doors: Sliding. End doors for staff use.

45	50	55	60	65	70
46	51	56	61	66	71
47	52	57	62	67	72
48	53	58	63	68	73
49	54	59	64	69	74

75	78	81	84	87	90
76	79	82	85	88	91
77	80	83	86	89	

CLASS B2K   2-SECTION UNITS

Built: 2002–03 by Bombardier Transportation, Bruges, Belgium.
Wheel Arrangement: B-2-B.   **Traction Motors:** Two Brush of 140 kW.
Seats: 52 (4).   **Weight:** 37 t.
Dimensions: 28.80 x 2.65 m.   **Braking:** Rheostatic.
Couplers: Scharfenberg.   **Maximum Speed:** 50 mph.
Doors: Sliding. End doors for staff use.

92	96	01	05	09	13
93	97	02	06	10	14
94	98	03	07	11	15
95	99	04	08	12	16

CLASS B07   2-SECTION UNITS

Built: 2007–10 by Bombardier Transportation, Bautzen, Germany.
Wheel Arrangement: B-2-B.   **Traction Motors:** Two Brush of 140 kW.
Seats: 52 (4).   **Weight:** 37 t.
Dimensions: **Braking:** Rheostatic.
Couplers: Scharfenberg.   **Maximum Speed:** 50 mph.
Doors: Sliding. End doors for staff use.

101	111	120	129	138	147
102	112	121	130	139	148
103	113	122	131	140	149
104	114	123	132	141	150
105	115	124	133	142	151
106	116	125	134	143	152
107	117	126	135	144	153
108	118	127	136	145	154
109	119	128	137	146	155
110					

6.3. EDINBURGH TRAMWAY

A new tramway is under construction in Edinburgh, however the scheme has been dogged by construction problems, with the originally planned terminus of Newhaven in the north of the city cut back to York Place in the city centre. In the west side of the city trams will run to Edinburgh Airport. At the time of writing the tramway is due to open in 2014, although it is likely that not all of the 27 trams on order will be required for the reduced system. The trams will be the longest to operate in the UK.

System: 750 V DC overhead.
Platform Height: 350 mm.
Depot & Workshops: Gogar.
Livery: White, red & black.

CAF   7-SECTION TRAMS

Built: 2009–11 by CAF, Irun, Spain.
Wheel Arrangement: Bo-Bo-2-Bo.   **Traction Motors:** 12 CAF of 80 kW.
Seats: 78.   **Weight:** 56.25 t.
Dimensions: 42.8 x 2.65 m.   **Braking:** Regenerative & electro hydraulic.
Couplers: Albert.   **Maximum Speed:** 50 mph.
Doors: Sliding plug.

251	257	263	268	273
252	258	264	269	274
253	259	265	270	275
254	260	266	271	276
255	261	267	272	277
256	262			

6.4. GLASGOW SUBWAY

This circular 4 ft gauge underground line is the smallest metro system in the UK, running for just over six miles. Operated by Strathclyde PTE the system has 15 stations. The entire passenger railway is underground, contained in twin tunnels, allowing for clockwise operation on the "outer" circle and anti-clockwise operation on the "inner" circle.

Trains are formed of 3-cars – either three power cars or two power cars sandwiching one of the newer trailer cars.

System: 600 V DC third rail.
Depot & Workshops: Broomloan.
Liveries: Strathclyde PTE carmine & cream unless stated.
S: New Subway livery (orange & grey).

SINGLE POWER CARS

Built: 1977–79 by Metro-Cammell, Birmingham. Refurbished 1993–95 by ABB Derby.
Wheel Arrangement: Bo-Bo.
Traction Motors: Four GEC G312AZ of 35.6 kW each.
Seats: 36.   **Dimensions:** 12.81 m x 2.34 m.
Couplers: Wedglock.   **Doors:** Sliding.
Weight: 19.6 t.   **Maximum Speed:** 33.5 mph.

101	**S**	108	**S**	115		122		128	
102		109		116	**S**	123		129	**S**
103	**S**	110		117	**S**	124		130	
104	**S**	111		118		125		131	
105	**S**	112		119	**S**	126		132	
106	**S**	113		120		127		133	
107		114		121					

INTERMEDIATE BOGIE TRAILERS

Built: 1992 by Hunslet Barclay, Kilmarnock.
Seats: 40.
Couplers: Wedglock.
Weight: 17.2t.
Dimensions: 12.70 m x 2.34 m.
Doors: Sliding.
Maximum Speed: 33.5 mph.
Advertising/Special liveries:

203 – Capital FM (blue & white).
204 – SPT Zonecard ticket (blue).
205 – Robert Burns (different images of the famous poet).

201	**S**	203	**AL**	205	**AL**	207		208
202		204	**AL**	206				

6.5. GREATER MANCHESTER METROLINK

Metrolink was the first modern tramway system in the UK, combining street running with longer distance running over former BR lines. The system opened in 1992 from Bury to Altrincham with a street section through the centre of Manchester and a spur to Piccadilly station. A second line opened in 2000 from Cornbrook to Eccles. A short spur off the Eccles line to MediaCityUK opened in September 2010 whilst the first part of the South Manchester Line to Chorlton and St Werburgh's Road opened in July 2011. In June 2012 the former National Rail line from Manchester to Oldham Mumps opened as a Metrolink line, extending the total route mileage to approximately 33 miles.

Further extensions are under construction as follows:

* Oldham Mumps–Rochdale station (on the the former National Rail line) due to open in late 2012. The line is also to be extended into Oldham and Rochdale town centres (2014).

* The East Manchester Line to Droylsden (late 2012) and Ashton-under-Lyne (winter 2013–14).

* The South Manchester Line from St Werburgh's Road to East Didsbury (on the fomer LMS line – summer 2013) and Manchester Airport (2016).

Further extensions will see trams running into Oldham and Rochdale town centres (2014) and a second city crossing in Manchester (by 2021).

Operator: RATP Dev.
System: 750 V DC overhead. High floor.
Depot & Workshops: Queens Road and Trafford.

T68 1000 SERIES   2-SECTION TRAMS

Built: 1991–92 by Firema, Italy. Chopper control.
Wheel Arrangement: Bo-2-Bo.
Dimensions: 29.0 x 2.65 m.
Doors: Sliding.
Weight: 45t.
Braking: Regenerative, disc and emergency track.
Traction Motors: Four GEC of 130 kW.
Seats: 82 (4).
Couplers: Scharfenberg.
Maximum Speed: 50 mph.

Liveries: White, dark grey & blue with light blue doors.
M: New Manchester Metrolink silver & yellow.

* Fitted with front-end valances, retractable couplers and controllable magnetic track brakes for on-street running mixed with private vehicles.

1001	*	(S)		1014	*	
1002	*			1015	*	(S)
1003	**M***			1016		
1004	*	Vans. The original since 1966 (S)		1017		
1005	*	(S)		1018		(S)
1006	*	Vans. The original since 1966 (S)		1019		(S)
1007	*	EAST LANCASHIRE RAILWAY		1020		LANCASHIRE FUSILIER
1008	*	(S)		1021	*	
1009	*			1022	*	POPPY APPEAL
1010	*	(S)		1023	*	
1011	*	Vans. The original since 1966 (S)		1024	*	
1012	*			1025	*	
1013	*			1026	*	

T68 2000 SERIES   2-SECTION TRAMS

Built: 1999 by Ansaldo, Italy. Chopper control. Fitted with front-end valances, retractable couplers and controllable magnetic track brakes for on-street running mixed with private vehicles.
Wheel Arrangement: Bo-2-Bo.   **Traction Motors:** Four GEC of 130 kW.
Dimensions: 29.0 x 2.65 m.   **Seats:** 82 (4).
Doors: Sliding.   **Couplers:** Scharfenberg.
Weight: 45t.   **Maximum Speed:** 50 mph.
Braking: Regenerative, disc and magnetic track.

Livery: White, dark grey & blue with light blue doors.

2001	2004
2002	2005
2003	2006

3000 SERIES   FLEXITY SWIFT   2-SECTION TRAMS

A total of 94 new Bombardier M5000 "Flexity Swift" currently being delivered to strengthen services on existing routes and for the extensions listed above. The last 32 cars listed are on order to replace all 32 of the T68 trams by early 2015.
Built: 2009–14 by Bombardier, Vienna, Austria.
Wheel Arrangement: Bo-2-Bo.
Traction Motors: Four Bombardier 3-phase asynchronous of 120 kW.
Dimensions: 28.4 x 2.65 m.   **Seats:** 52.
Doors: Sliding.   **Couplers:** Scharfenberg.
Weight: 39.7 t.   **Maximum Speed:** 50 mph.
Braking: Regenerative, disc and magnetic track.

Livery: New Manchester Metrolink silver & yellow.

3001	3020	3039	3058	3077
3002	3021	3040	3059	3078
3003	3022	3041	3060	3079
3004	3023	3042	3061	3080
3005	3024	3043	3062	3081
3006	3025	3044	3063	3082
3007	3026	3045	3064	3083
3008	3027	3046	3065	3084
3009	3028	3047	3066	3085
3010	3029	3048	3067	3086
3011	3030	3049	3068	3087
3012	3031	3050	3069	3088
3013	3032	3051	3070	3089
3014	3033	3052	3071	3090
3015	3034	3053	3072	3091
3016	3035	3054	3073	3092
3017	3036	3055	3074	3093
3018	3037	3056	3075	3094
3019	3038	3057	3076	

6.6. LONDON TRAMLINK

This system runs through central Croydon via a one-way loop, with lines radiating out to Wimbledon, New Addington and Beckenham Junction/ Elmers End, the total route mileage being 18½ miles. It opened in 2000 and is now operated by Transport for London. Six new Stadler trams entered traffic in spring 2012.

System: 750 V DC overhead. **Platform Height:** 350 mm.
Depot & Workshops: Therapia Lane, Croydon.
Livery: Light grey & lime green with a blue solebar.
Advertising livery:
2554 – Love Croydon (purple & blue)

BOMBARDIER   3-SECTION TRAMS

Built: 1998–99 by Bombardier, Vienna, Austria.
Wheel Arrangement: Bo-2-Bo. **Traction Motors:** Four of 120 kW each.
Dimensions: 30.1 x 2.65 m. **Seats:** 70.
Doors: Sliding plug. **Couplers:** Scharfenberg.
Weight: 36.3t. **Maximum Speed:** 50 mph.
Braking: Disc, regenerative and magnetic track.

2530	2534	2538	2542	2546	2550
2531	2535	2539	2543	2547	2551
2532	2536	2540	2544	2548	2552
2533	2537	2541	2545	2549	2553

Name: 2535 STEPHEN PARASCANDOLO 1980–2007

STADLER   5-SECTION TRAMS

Six new Variobahn trams entered traffic in 2012.
Built: 2011–12 by Stadler, Berlin, Germany.

Wheel Arrangement:	**Traction Motors:** 8 of 45 kW.
Dimensions: 32.4 x 2.65 m.	**Seats:** 72.
Doors: Sliding plug.	**Couplers: Albert.**
Weight: 41.5 t.	**Maximum Speed:** 50 mph.
Braking:	

2554 **AL**	2555	2556	2557	2558	2559

6.7. NOTTINGHAM EXPRESS TRANSIT

This light rail system opened in 2004. Line 1 runs for 8¾ miles from Station Street, Nottingham (alongside Nottingham station) to Hucknall, including a short spur to Phoenix Park. There is around three miles of street running through Nottingham. Extensions are planned to open in 2014 to Clifton (Line 2) to the south of Nottingham, and Chilwell via Beeston to the west (Line 3). 22 Alstom Citadis trams are on order for these extensions.

The system is operated by the Tramlink Nottingham consortium (formed of Taylor Woodrow, Alstom, Keolis, Wellglade, Meridiam Infrastructure and Infravia Fund).

System: 750 V DC overhead.   **Platform Height:** 350 mm.
Depot & Workshops: Wilkinson Street.

BOMBARDIER INCENTRO   5-SECTION TRAMS

Built: 2002–03 by Bombardier, Derby Litchurch Lane Works.

Wheel Arrangement: Bo-2-Bo.	**Traction Motors:** 8 x 45 kW wheelmotors.
Dimensions: 33.0 x 2.4 m	**Seats:** 54 (4).
Doors: Sliding plug.	**Couplers:** Not equipped.
Weight: 36.7 t.	**Maximum Speed:** 50 mph.

Braking: Disc, regenerative and magnetic track for emergency use.
Standard livery: Black, silver & green unless stated.
Advertising liveries:

201 – Nottinghamcontemporary.org (yellow/light blue & white).
204 – Glide into Nottingham (blue & white).
208 – e.on (red).
211 – Diamond Jubilee (red).

201	**AL**	Torvill and Dean	209		Sid Standard
202		DH Lawrence	210		Sir Jesse Boot
203		Bendigo Thompson	211	**AL**	Robin Hood
204	**AL**	Erica Beardsmore	212		William Booth
205		Lord Byron	213		Mary Potter
206		Angela Alcock	214		Dennis McCarthy
207		Mavis Worthington	215		Brian Clough
208	**AL**	Dinah Minton			

6.8. MIDLAND METRO

This system opened in 1999 and has one 12½ mile line from Birmingham Snow Hill to Wolverhampton along the former GWR line to Wolverhampton Low Level. On the approach to Wolverhampton it leaves the former railway alignment to run on-street to the St George's terminus. It is operated by Travel West Midlands Ltd. An extension is under construction from Birmingham Snow Hill into the city centre (to terminate near New Street station) and a city centre loop in Wolverhampton is also planned. 25 CAF Urbos 3 trams are on order to replace the Ansaldo trams and to serve the expanded system.

System: 750 V DC overhead.   **Platform Height:** 350 mm.
Depot & Workshops: Wednesbury.

ANSALDO   2-SECTION TRAMS

Built: 1998–99 by Ansaldo Transporti, Italy.
Wheel Arrangement: Bo-2-Bo.   **Traction Motors:** Four of 105 kW each.
Dimensions: 24.00 x 2.65 m.   **Seats:** 52.
Doors: Sliding plug.   **Couplers:** Not equipped.
Weight: 35.6 t.   **Maximum Speed:** 43 mph.
Braking: Regenerative, disc and magnetic track.

Standard livery: Dark blue & light grey with green stripe, yellow doors & red front end.

MW: New Network West Midlands tram livery (silver & pink).
Note: 01 and 02 have been withdrawn and are stored at Wednesbury depot.

01	(S)		09	MW	JEFF ASTLE
02	(S)		10	MW	JOHN STANLEY WEBB
03		RAY LEWIS	11		THERESA STEWART
04		SIR FRANK WHITTLE	12		
05	MW	SISTER DORA	13		ANTHONY NOLAN
06		ALAN GARNER	14		JIM EAMES
07	MW	BILLY WRIGHT	15		AGENORIA
08		JOSEPH CHAMBERLAIN	16		GERWYN JOHN

6.9. SHEFFIELD SUPERTRAM

This system opened in 1994 and has three lines radiating from Sheffield City Centre. These run to Halfway in the south-east, with a spur from Gleadless Townend to Herdings Park, to Middlewood in the north with a spur from Hillsborough to Malin Bridge and to Meadowhall Interchange in the north east, adjacent to the large shopping complex. The total length is 18 miles. The system is a mixture of on-street and segregated running.

The cars are owned by South Yorkshire Light Rail Ltd, a subsidiary of South Yorkshire PTE. The operating company, South Yorkshire Supertram Ltd, is contracted to Stagecoach who operate them as Stagecoach Supertram.

Because of severe gradients in Sheffield (up to 1 in 10) all axles are powered on the vehicles, which have low-floor outer sections.

System: 750 V DC overhead.   **Platform Height:** 450 mm.
Depot & Workshops: Nunnery.
Standard livery: Stagecoach (All over blue with red & orange ends).
Non-standard/Advertising liveries: 111 – East Midlands Trains (blue).
116 – Genting Club (black & red).
120 – Original Sheffield Corporation tram livery (cream & blue).

SIEMENS   3-SECTION TRAMS

Built: 1993–94 by Siemens, Krefeld, Germany.
Wheel Arrangement: B-B-B-B.
Traction Motors: Four monomotor drives of 250 kW.
Dimensions: 34.75 x 2.65 m.   **Seats:** 80 (6).
Doors: Sliding plug.   **Couplers:** Not equipped.
Weight: 52t.   **Maximum Speed:** 50 mph.
Braking: Regenerative, disc and emergency track.

101	106	110		114		118		122
102	107	111	**AL**	115		119		123
103	108	112		116	**AL**	120	**0**	124
104	109	113		117		121		125
105								

6.10. TYNE & WEAR METRO

The Tyne & Wear Metro system covers 48 route miles and can be described as the UK's first modern light rail system.

The initial network opened between 1980 and 1984, consisting of a line from South Shields via Gateshead and Newcastle Central to Bank Foot (extended to Newcastle Airport in 1991) and the North Tyneside loop (over former BR lines) serving Tynemouth and Whitley Bay with a terminus at St James. A more recent extension was from Pelaw to Sunderland and South Hylton in 2002, using existing heavy rail infrastructure between Heworth and Sunderland.

The system is owned by Nexus (the Tyne & Wear PTE) and operated by DB Regio.

System: 1500 V DC overhead.   **Depot & Workshops:** South Gosforth.

METRO-CAMMELL   2-SECTION UNITS

Built: 1978–81 by Metropolitan Cammell, Birmingham (Prototype cars 4001 and 4002 were built by Metropolitan Cammell in 1975 and rebuilt 1984–87 by Hunslet TPL, Leeds). A further rebuild is now taking place at Wabtec, Doncaster. This commenced in 2010 and is due to be complete in 2015.
Wheel Arrangement: B-2-B.
Traction Motors: Two Siemens of 187 kW each.
Dimensions: 27.80 x 2.65 m.   **Seats:** 68 (**TW** = 64).
Doors: Sliding plug.   **Couplers:** BSI.
Weight: 39.0 t.   **Maximum Speed:** 50 mph.
Braking: Air/electro magnetic emergency track.

Standard livery: Red & yellow unless otherwise indicated.
B Blue & yellow.   **G** Green & yellow.
0 (4001) Original 1975 Tyne & Wear Metro livery of yellow & cream.
0 (4027) Original North Eastern Railway style (red & white).
TW New Tyne & Wear Metro (grey, black & yellow).

Advertising liveries:

4002 – Tyne & Wear Metro (orange & black).
4020 – Modern Apprenticeships (white, red & blue).
4038 – Talktofrank.com (white).
4040 – Cut your CO_2 day (blue & white).
4045 – Newcastle International Airport – 75 years (purple).
4055 – European Regional Development Fund (blue & yellow).
4067 – Artist Alexander Millar (white & blue).
4075 – Tyne & Wear Public Services (purple & white).
4080 – South Shields market (white).
4084 – Pop smartcard (various).

4001	**0**	4016	**B**	4031	**B**	4046	**TW**	4061	**TW**	4076	**B**
4002	**AL**	4017		4032		4047	**B**	4062	**G**	4077	
4003		4018	**G**	4033		4048		4063		4078	
4004	**G**	4019		4034		4049	**TW**	4064		4079	
4005		4020	**AL**	4035	**B**	4050		4065		4080	**AL**
4006		4021		4036	**G**	4051	**G**	4066	**B**	4081	**B**
4007		4022	**TW**	4037		4052		4067	**AL**	4082	**TW**
4008		4023	**G**	4038	**AL**	4053	**B**	4068		4083	**B**
4009		4024	**B**	4039	**B**	4054	**B**	4069		4084	**AL**
4010		4025		4040	**AL**	4055	**AL**	4070		4085	
4011		4026		4041	**TW**	4056		4071		4086	
4012		4027	**0**	4042	**TW**	4057	**TW**	4072	**TW**	4087	
4013		4028		4043		4058	**TW**	4073		4088	
4014		4029	**B**	4044		4059		4074	**TW**	4089	
4015		4030		4045	**AL**	4060		4075	**AL**	4090	**TW**

Names:

4026	George Stephenson	4073	Danny Marshall
4060	Thomas Bewick	4077	Robert Stephenson
4064	Michael Campbell	4078	Ellen Wilkinson
4065	DAME Catherine Cookson		

7. CODES

7.1. LIVERY CODES

Code Description

1　"One" (metallic grey with a broad black bodyside stripe. White National Express "interim" stripe as branding).

AL　Advertising/promotional livery (see class heading for details).

B　BR blue.

CN　Connex/Southeastern (white with black window surrounds & grey lower band).

CX　Connex (white with yellow lower body & blue solebar).

EU　Eurostar (white with dark blue & yellow stripes).

FB　First Group dark blue.

FU　First Group "Urban Lights" (varying blue with pink, white & blue markings on the lower bodyside).

GA　Greater Anglia (white with red doors).

GV　Gatwick Express EMU (red, white & indigo blue with mauve & blue doors).

HC　Heathrow Connect (grey with a broad deep blue bodyside band & orange doors).

HE　Heathrow Express (grey & indigo with blue or purple doors and black window surrounds).

LM　London Midland (grey & green with broad black stripe around the windows).

LO　London Overground (all over white with a blue solebar & black window surrounds).

LT　London Transport maroon & cream.

ME　Merseyrail (metallic silver with yellow doors).

N　BR Network SouthEast (white & blue with red lower bodyside stripe, grey solebar & cab ends).

NC　National Express white (white with blue doors).

NO　Northern (deep blue, purple & white).

NX　National Express (white with grey ends).

RK　Railtrack (green & blue).

RM　Royal Mail (red with yellow stripes above solebar).

RR　Regional Railways (dark blue/grey with light blue & white stripes, three narrow dark blue stripes at cab ends).

SB　Southeastern High Speed (all over blue with black window surrounds).

SC　Strathclyde PTE (carmine & cream lined out in black & gold).

SD　South West Trains outer suburban livery {Class 450 style} (deep blue with red doors & orange & red cab sides).

SE　Southeastern (all over white with black window surrounds, light blue doors and (on some units) dark blue lower bodyside stripe).

SL　Silverlink (indigo blue with a white stripe, green lower body and yellow doors).

SN　Southern (white & dark green with light green semi-circles at one end of each vehicle. Light grey band at solebar level).

SP　Strathclyde PTE {revised} (carmine & cream, with a turquoise stripe).

SR	ScotRail – Scotland's Railways (dark blue with Scottish Saltire flag & white/light blue flashes).
SS	South West Trains inner suburban {Class 455} (red with blue & orange flashes at unit ends).
ST	Stagecoach {long-distance stock} (white & dark blue with dark blue window surrounds and red & orange swishes at unit ends).
VT	Virgin Trains silver (silver, with black window surrounds, white cantrail stripe & red roof. Red swept down at unit ends. Black & white striped doors).
YR	West Yorkshire PTE/Northern EMUs (red, lilac & grey).

7.2. OWNER CODES

A	Angel Trains
BN	Beacon Rail Leasing
BT	Bombardier Transportation UK
E	Eversholt Rail (UK)
EM	East Midlands Trains
EU	Eurostar (UK)
FC	First Capital Connect
HE	British Airports Authority
LY	Lloyds Banking Group
P	Porterbrook Leasing Company
QW	QW Rail Leasing
RM	Royal Mail
SB	SNCB/NMBS (Société Nationale des Chemins de fer Belges Nationale Maatschappij der Belgische Spoorwegen)
SF	SNCF (Société Nationale des Chemins de fer Français)
SW	South West Trains
VI	Virgin Trains
WC	West Coast Railway Company

7.3. OPERATOR CODES

C2	c2c
DB	DB Schenker
EA	Greater Anglia
EU	Eurostar (UK)
FC	First Capital Connect
HC	Heathrow Connect
HE	Heathrow Express
LM	London Midland
LO	London Overground
ME	Merseyrail
NO	Northern
SE	Southeastern
SN	Southern
SR	ScotRail
SW	South West Trains
VW	Virgin Trains

7.4. ALLOCATION & LOCATION CODES

Code	Location	Depot Operator
AD	Ashford	Hitachi
AF	Ashford Chart Leacon Works	Bombardier Transportation
BD	Birkenhead North	Merseyrail
BF	Bedford Cauldwell Walk	First Capital Connect
BI	Brighton Lovers Walk	Southern
BM	Bournemouth	South West Trains
CE	Crewe International	DB Schenker Rail (UK)
CS	Carnforth	West Coast Railway Company
DY	Derby Etches Park	East Midlands Trains
EM	East Ham (London)	c2c
FF	Forest (Brussels)	SNCB/NMBS
GW	Glasgow Shields Road	ScotRail
HE	Hornsey (London)	First Capital Connect
IL	Ilford (London)	Greater Anglia
LB	Loughborough Works	Brush Traction
LG	Longsight (Manchester)	Northern
LM	Long Marston (Warwickshire)	Motorail Logistics
LY	Le Landy (Paris)	SNCF
MA	Manchester Longsight	Alstom
NG	New Cross Gate (London)	London Overground
NL	Neville Hill (Leeds)	East Midlands Trains/Northern
NN	Northampton King's Heath	Siemens
NT	Northam (Southampton)	Siemens
OH	Old Oak Common Heathrow (London)	Heathrow Express
RM	Ramsgate	Southeastern
RY	Ryde (Isle of Wight)	South West Trains
SG	Slade Green (London)	Southeastern
SL	Stewarts Lane (London)	Southern/Orient Express Hotels
SO	Soho (Birmingham)	London Midland
SU	Selhurst (Croydon)	Southern
TI	Temple Mills (London)	Eurostar
WB	Wembley (London)	Alstom
WD	Wimbledon (London)	South West Trains
ZA	RTC Business Park (Derby)	Railway Vehicle Engineering
ZB	Doncaster Works	Wabtec Rail
ZC	Crewe Works	Bombardier Transportation
ZD	Derby Works	Bombardier Transportation
ZG	Eastleigh Works	Arlington Fleet Services
ZH	Springburn Depot (Glasgow)	Railcare
ZI	Ilford Works	Bombardier Transportation
ZJ	Stoke Works	Axiom Rail
ZK	Kilmarnock Works	Wabtec Rail Scotland
ZN	Wolverton Works	Railcare
ZR	York (Holgate Works)	Network Rail

*= unofficial code.